Recycling Projects for the Evil Genius™

Evil Genius™ Series

Bike, Scooter, and Chopper Projects for the Evil Genius

Bionics for the Evil Genius: 25 Build-It-Yourself Projects

Electronic Circuits for the Evil Genius, Second Edition: 64 Lessons with Projects

Electronic Gadgets for the Evil Genius: 28 Build-It-Yourself Projects

Electronic Sensors for the Evil Genius: 54 Electrifying Projects

50 Awesome Auto Projects for the Evil Genius

50 Green Projects for the Evil Genius

50 Model Rocket Projects for the Evil Genius

51 High-Tech Practical Jokes for the Evil Genius

46 Science Fair Projects for the Evil Genius

Fuel Cell Projects for the Evil Genius

Holography Projects for the Evil Genius

Mechatronics for the Evil Genius: 25 Build-It-Yourself Projects

Mind Performance Projects for the Evil Genius: 19 Brain-Bending Bio Hacks

MORE Electronic Gadgets for the Evil Genius: 40 NEW Build-It-Yourself Projects

101 Outer Space Projects for the Evil Genius

101 Spy Gadgets for the Evil Genius

125 Physics Projects for the Evil Genius

123 PIC® Microcontroller Experiments for the Evil Genius

123 Robotics Experiments for the Evil Genius

PC Mods for the Evil Genius: 25 Custom Builds to Turbocharge Your Computer

PICAXE Microcontroller Projects for the Evil Genius

Programming Video Games for the Evil Genius

Recycling Projects for the Evil Genius

Solar Energy Projects for the Evil Genius

Telephone Projects for the Evil Genius

30 Arduino Projects for the Evil Genius

25 Home Automation Projects for the Evil Genius

22 Radio and Receiver Projects for the Evil Genius

Recycling Projects for the Evil Genius™

Russel Gehrke

New York Chicago San Francisco Lisbon London Madrid
Mexico City Milan New Delhi San Juan Seoul
Singapore Sydney Toronto

The McGraw-Hill Companies

McGraw-Hill books are available at special quantity discounts to use as premiums and sales promotions, or for use in corporate training programs. To contact a representative, please e-mail us at bulksales@mcgraw-hill.com.

Recycling Projects for the Evil Genius™

The pages within this book were printed on acid-free paper containing 100% post-consumer fiber.

1 2 3 4 5 6 7 8 9 0 WDQ/WDQ 1 6 5 4 3 2 1 0

ISBN 978-0-07-173612-1
MHID 0-07-173612-3

Sponsoring Editor
Judy Bass

Editorial Supervisor
Stephen M. Smith

Production Supervisor
Richard C. Ruzycka

Acquisitions Coordinator
Michael Mulcahy

Project Manager
Patricia Wallenburg, TypeWriting

Copy Editor
James Madru

Proofreader
Paul Tyler

Indexer
Karin Arrigoni

Art Director, Cover
Jeff Weeks

Composition
TypeWriting

This book is dedicated to the memory of the late governor
from the state of Missouri, Mel Carnahan, and to his family.

About the Author

Russel Gehrke is an engineer and inventor well-known in alternative energy circles. He is a professional consultant and chief technical officer for Agrifuels, LLC, and has been featured on the Discovery Channel and in several episodes of the syndicated television show *Coolfuel Roadtrip*, his own pilot TV show, *The Eco Outlaws*, and the History Channel's *Modern Marvels*. Mr. Gehrke has built hot rods for Willie Nelson and custom motorcycles for John Paul DeJoria, Merle Haggard, and others. He is the inventor of "The Green Box," a patent-pending technology that combines multiple sources of affordable renewable energies to produce electricity for the home owner in a more cost-effective way.

Contents

Foreword

How lucky am I to have an Evil Genius on speed dial? It is like being able to drive a truck over your favorite superhero for laughs. That superhero may wince but the impact will not faze him. Here is an excerpt from a recent phone call I had with the Evil Genius himself, Russel Jay Gehrke: "Hey Russel, I think I want to buy a generator in case an earthquake strikes and wipes out my house power." Without a beat Russel responds, "Well, Phil, sounds like a plan but wouldn't it be cool if you had a mobile generator operating as a semi-permanent power source for the house to lessen your normal energy use? I'll send you a photo after I whip it up." This is why I am lucky to have Russel on speed dial.

Russel turns normal questions into challenges to create practical tools for everyday use. He has a firm understanding of our current technologies and applications as well as their limitations. His Evil Genius will help us break new ground in our evolution and this book is proof of that.

Over the last ten years, I have been living and working in Los Angeles as a producer on *CSI: Crime Scene Investigation*. A few seasons ago, I rounded up a team of biodiesel professionals to help me transition *CSI*'s trucks and generators to run on the domestic and renewable fuel. One of the key players on the team was Russel. He is a humble man from southern Missouri with a multitude of specialties, one of which is biodiesel production. I value his expertise because I certainly did not want any power problems due to fuel issues. Russel made sure each engine ran smoothly and on 100 percent biodiesel. Today more shows are running on biodiesel and making strides to more-sustainable production practices.

When Russel sent me the picture of his newest invention, my grin spread as it loaded on my smart phone's screen. It was a metal box topped with a solar panel. Fastened to the side of the box was a small wind turbine mounted on a telescopic pole. Inside the box was a small diesel generator bolted beside an isolated battery array. The "Green Box," as he calls it, is a multi-equipped hybrid generator capable of producing and storing power. The Evil Genius has bounced back once again like the superhero he is with an answer to my challenge. Russel, let me thank you again for writing this incredible book, and I want to place a order for my own Green Box...sorry, Sears.

Phil Conserva
TV Producer

Acknowledgments

Writing a book is a frightful thing for me, because when you write a book you are basically opening up your head and heart, letting pure strangers see the world through your own eyes. This book is full of personal projects I've done for the past twenty-plus years; some worked great while others worked not so great. The benefit of failure is that you can share with others what not to do, and that is the mark of a good how-to book.

First, I want to thank the people who made this book "a book"—taking my fractured thoughts and directions and mixing them together in a manner that sees beyond my own rose-colored glasses—sponsoring editor Judy Bass, project manager Patricia Wallenburg, and copy editor James Madru, who had the tough job of breaking the Gehrke code I so often write in. The McGraw-Hill gang keeps me focused and motivated. I must also thank my old friend Carl Vogel, the author of *Build Your Own Electric Motorcycle*, for helping McGraw-Hill find me so I could write this book and also my first book, *Renewable Energies for Your Home*. Another big help in getting things done was Tamera S. Wright; she did all the sketches in this book and was always a positive light for me at just the right times. I also must mention Kristina Fitzsimmons, who helps me with my press and web site, and makes things move forward beyond my own expectations.

Second, I want to thank all my friends in the State of Missouri's Department of Natural Resources for guidance over the past twenty years. I also need to thank the Environmental Improvement and Energy Resources Authority for its help and advice, and specifically Kristin Allan Tipton and Lee Fox, for always taking my calls and taking the time to let me pick their brains no matter how crazy my idea was. The cornerstone for my larger projects over the years has been a fella named Jimmy Story; he's the Business Program Manager for a group called Missouri Enterprise that helps Missouri businesses like mine. And I want to thank the bunch at Agrifuels, LLC—Paul, Kathy, Jason, and Matt Hoar as well as Rich Sulinski—and my friend Phil Conserva, for all his great ideas and for always being on the lookout for a new project for me to work on with him.

To all the others and my family who contributed immeasurably to this book I offer my thanks, and forgive me for not listing your names. Last but not least, I want to thank *you* for taking the time to read this book, and I hope you find value in what was put forth in its pages.

Russel Gehrke

CHAPTER 1

My Shade of Green

When it comes to recycling, it's the *shade* of green that is important, and we all have a choice about what shade of green we live by. The very title of this book may give you the mistaken impression that I'm just another stereotypical "eco guy." Trust me, most people think that same thing when they first meet me. I don't want people to feel the need to blurt out, "I recycle too!"; hide their disposable plastic bottle of water; or cancel their order for rare steak, nor do I want to watch them scan the room for the nearest exit. Once people get to know me, they realize that I'm not going to chastise them for driving a gas guzzler or cry over the loss of the bugs that are splattered on their car's windshield. I also happily make it a point to keep my vocabulary devoid of the word *dude* at the beginning of every sentence. I'm just simply in favor of more efficient and effective ways to put as many of our natural resources to use as possible that won't require us to send our much-needed money overseas to help fund others who may not have our own or their own populations' best interests at heart. We won't ever be able to effectively change anything with scary doomsday predictions, politically driven actions, or half-baked solutions that don't even help us to sleep better at night. Should we all drink the Kool-Aid just because some celebrity tells us that it is the right thing to do? If we could just for a moment pull all our hands out of the proverbial cookie jar, stop with the politics and the anti-industry, global warming, end-of-the-world speak, and realize that our need is to focus on the things we ourselves can do to change our own shade of green, we would be taking one small step in the right direction.

Tell Me About the Good Ol' Days

My grandfather lived to be 100 years old. He was born on April 16, 1900, in the state of Kansas. He lived through a handful of wars, the Great Depression, and the dust bowl and saw the world go from candlelight to the Internet. In his lifetime, the event that most affected his well-being and lifestyle was not the cold war, civil rights, or humans landing on the moon. It was the loss of resources caused by the Great Depression and the dust bowl over a half century ago that had made so many in the Midwest fight for their very survival. I remember those stories, and as a kid, they made me aware of the natural world and how quickly things can change. What we look at as the green movement today was a matter of survival back then, and the lesson to be learned from my grandfather's stories is that at any moment our world could upheave, and we all could be in the same shoes as he was. Thus, recycling to me is a form of risk management. The idea of being able to recycle the things I've made or bought for my family and reuse them as something else is very rewarding, and it becomes, in essence, another source of income. The green shade of money is a very nice thing.

Out of the Box

When I was just a little Russ, about eight years old or so, my dad would haul our trash to the landfill in the back of his truck on the way to town. My brothers and I went along for the ride because we loved to watch the track loader push and pile the trash. I loved to spot things that I thought I could fix, and Dad would let me take some things home, such as old record players or TVs; anything electronic, I loved. I was one of those "fall between the cracks" kids who really couldn't read or write, but for some reason, I just had a grasp on electrical circuits and mechanical things. At the end of that same summer, I went to a fair and was inspired by the rides. When I got home, I made my own amusement ride from an aluminum ladder, a toy seat, and a block and tackle setup. I played Beatles music to accompany the "ride" on a tape recorder I had fixed that also came from the dump. I charged my friends a quarter for each ride, and that was my first taste of entrepreneurship. I was just so excited about finding what people tossed away; their trash was my treasure. I may not have been able to read or write at the time like my eight-year-old peers, but I knew more about electronic and mechanical things than most adults. In a few short years I learned how to deal with my learning disability and have never really looked back or felt like I had been cheated somehow. Being dyslexic had turned out to be a gift—one that forced me to be more independent, learn on my own terms, in my own way, and at my own speed. In short, I learned how to think outside the box. Seeing the world around me differently from those who tried to teach me is what has made me such a creative problem solver and has served me well as an adult. Identifying the problems won't solve them, but it's hard to solve a problem unless it's identified, and seeing a problem is easier if you stay outside the box. As with most things in life, it's better to stay out the box than to try to get out of it once you're in it.

It's a Brick House

In the early 1990s, I started working with my family on what I call a "real job"—recycling plastic into plastic lumber. It was very difficult to make a living from this business because the price of plastic was high, and the materials we produced had to compete with treated wooden lumber, which was cheaper at the time. We knew we had to do something different just to keep our doors open and survive. Then one day while I was driving across town, I saw some Amish men working on a roof, and that gave me an idea. The same machines that we used to make plastic lumber might allow us to mix asphalt-based roofing tear-off (old roofing that was being replaced) into our plastic and cut our manufacturing cost. It worked, and we were on a new path.

The best part about using waste roofing shingles was that we would get paid to recycle the old roofing. The landfill got paid to dispose of such material, and we were by no means a free dump. In fact, the demand was so great for our service that we stopped recycling plastics and started making products from 100 percent asphalt roofing waste. We made bricks, blocks, pavers, cold patch for potholes, and even a hot-mix asphalt additive that cut material cost up to 40 percent for asphalt plants. I even built a house using the recycled asphalt blocks some 11 years ago, and that house still looks as good as it did when it was new (Figures 1-1 and 1-2).

In 2000, a new project came to life for our family business with the help of Missouri Enterprises, the University of Missouri Rolla, and the Lemay Center for Composite Technologies. The idea for this new project was based on the

Figure 1-1

Figure 1-2

integration of recycled materials into engineered composites to provide potential cost savings on virgin materials and to provide a solution to the ever-growing problem of solid-waste disposal and recycling costs. Roofing shingles contain petroleum-based binders and fillers that are a valuable resource in composite production. Composites offer inherent advantages over traditional materials in terms of corrosion resistance, design flexibility, and extended service life. Scrap roofing shingles can be used as a core material in glass fiber–reinforced composite materials that offer potentially low-cost solutions for sound barrier systems, railroad ties, and other building materials, including blocks. To accomplish this, processes have been developed for shredding scrap roof shingles, for making shingle blocks, and for filling hollow composite tubes.

Mechanical testing was performed to compare the performance of filled composite tubes with that of hollow tubes and oak beams. Filled tubes were shown to have improved ultimate flexible strength by preventing buckling and crushing. Tests also were conducted to evaluate the sound-attenuating capability of recycled-shingle composite walls. It was shown that the mean sound level at the backside of the wall, measured in decibels, was greatly reduced, indicating the potential of recycled shingles in a sound barrier system. Reuse of scrap asphalt roof shingles can reduce problems of landfill disposal, and the material can be used for producing blocks, railroad ties, decorative arrangements, and sound barriers. Railroad ties traditionally have been made of wood, but recent research has focused on alternate designs and materials. Although inexpensive, wood has many disadvantages that make the use of alternative materials feasible and more affordable over the long term. The disadvantages of wood include rot, insect attack, tie plate cutting, failure by degradation of mechanical properties, and future lumber restrictions and regulations. Wood railroad ties are soaked in creosote to protect them from environmental conditions, such as insects and rot. Creosote is a hazardous material known to cause cancer, and such ties must be disposed of safely when they are removed, adding to the cost. In addition, over time the creosote begins to seep out of railroad ties, causing damage to the environment. Several research groups have designed new railroad ties using materials such as plastic-wood composites and commingled plastics and reinforced concrete. Some research has focused on improving the mechanical properties of railroad ties through the use of a composite wrap. To minimize cost, a parametric study was performed to minimize the volume of the composite wrapping while maintaining improved mechanical properties. A wooden railroad tie partially covered with glass fiber–reinforced plastic was modeled, tested, and found to have improved mechanical properties when compared with traditional wooden ties. Although strengthening a railroad tie will result in fewer failures owing to fatigue, other failure modes were not addressed. It is expected that rot, insect attack, and creosote seepage will remain major factors in the failure of wooden railroad ties.

A recycled shingle–core composite railroad tie is not susceptible to rot or attack from fungi or insects because of its inability to absorb moisture and the absence of protein or other easily decomposed material in its composition. The composite tie consists of two components, the recycled shingle core and the composite material covering. The composite covering will bear the load on the structure, whereas the shingle material forms the low-cost core. At present, two stages of research have been completed. In the first stage, small-scale samples (4 × 4 × 40 inches) of filled composite tubes were fabricated and tested. Flexure tests were performed to evaluate the performance of shingle-core composite tubes. Also, the sound-attenuating capabilities of the filled tubes were investigated to see if they would work as a sound barrier along busy highways if made into a wall. The stage involved identifying a practical method of manufacturing composite-wrapped recycled shingle–core railroad ties, and further research was performed to evaluate the feasibility of applying the composite wrap to the recycled shingle core by hand. Thus 4 × 8 × 40 inch samples were manufactured by hand layup, and three-point flexure tests were performed (Figures 1-3 and 1-4).

The tests showed that the composite railroad tie could withstand greater loads and have a longer life span than conventional creosote-treated oak ties. To make this long story short, we had to shut our doors after about a 10-year run. The project no longer exists owing to competition, personal matters, and a lack of funding. To this day, there are two asphalt plants in our area that recycle thousands of tons of roofing waste into hot mix. Thus it's nice to see what we started still goes on and is successful (Figure 1-5).

Paper or Plastic?

When I'm in the checkout line at the grocery store and am asked if I want paper or plastic, I say, "Plastic, please," because I can use those pesky plastic bags for so many useful things. I do hate to see folks just toss them away or, worse, let them loose as litter to end up as trash caught in trees,

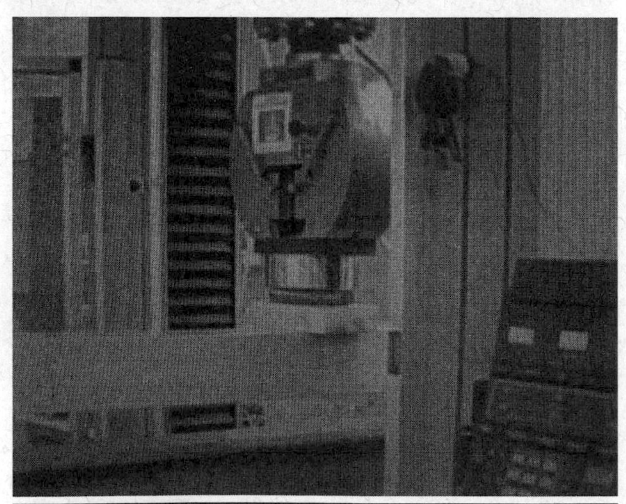

Figure 1-3

Figure 1-4

Figure 1-5

floating in waterways, or blowing around just about everywhere I look. I like to take those bags and make what I call a "crispy treat." The idea came to me in early 1990 to take a filler such as sand, crushed glass, or aggregate; heat it up to a set temperature; and then start putting plastic bags in with it and stir (in the manufacturing plant I used a large heated mixer). The plastic, once melted, sticks or binds the filler (aggregate) together, making what looks like a concrete marshmallow "crispy treat." I then place this "crispy treat" in a mold to form it into a brick, block, or whatever. It can take 100 bags to make one block, so you can see how you'll be looking for bags if you get into making "crispy treat" blocks.

You'll get to where you'll start asking family members, neighbors, and friends to save their bags for you. This is the whole idea about recycling in your own home. As you become more successful at making these blocks, you can use them at home or sell them to make extra cash, and you'll begin to need more plastic bags as a resource. When you add value to a resource such as a used plastic bag at your home, that's great because it didn't travel halfway across the country to be recycled into whatever. You just made a greener choice and saved a lot of energy and money, both of which make recycling worthwhile as well as put money back in your pocket.

Safer Chemicals

I've always loved chemistry because it fits well with the way I like to think outside the box. Chemicals are basically parts of a machine that you can build by adding or taking away certain parts to do a specific job or create a predictable outcome. Things we use in our homes today can be very dangerous because of the types and percentages of the chemicals in them. One of the major reasons our household chemicals have gotten more dangerous is consumer demand. We as consumers ask for and purchase things for our homes that promise at *first glance* to do things faster, cheaper, and better. The manufacturers of these goods have met our demands and given us what we want. In so doing, they have created stronger and more reactive chemicals that we use in our environments. At present, there is a consumer demand calling for products that are more ecological, green, and earth-friendly. Because of successful marketing and "greenwashing," many of these products are just plain costing us more and often don't do a good job.

Mean Green

Today, it seems that the word *green* is passed around far too often and is found everywhere we look. In the news, the speeches of our political leaders, and the stores in which we shop, *green* is the buzzword often used to get into our hearts, heads, and pocketbooks. I feel as though I'm being brainwashed (or "greenwashed") by all the green talk and green hype. I predict that in the near future we'll hear the word *greenwashed* used by the uber green folks to point fingers at companies and even our political leaders who use the word *green* to sell more goods or to push their agendas and policies. The same is true with all those products that say that they are earth-friendly but, again, cost more.

I enjoy making my own cleaners, and more often than not, I save money doing it. We all can easily find baking soda, vinegar, alcohol, vegetable oil, and even rainwater. These ingredients, along with a few special others, can do a better and safer job of cleaning and offering protection around your home. Examples include a chalk line that ants won't cross and a super degreaser from waste vegetable oil that can be used to make furniture polish, laundry stain remover, or a bug/tar remover for your car.

My Shade of Green

My hope for all of us is that we are able to trust ourselves and not be scared that we have to change our lives drastically just to satisfy the people who want to make us feel guilty because some cows pass gas. It is my opinion that we all can become better conservationists and more effective consumers without having to dispose of the many comforts of life to which we have become accustomed. I also want people to take a step back and look at all the information available to us on the environment with a critical eye. Question those dramatic scenarios, and just look for the facts in all of it. The most destructive thing that we have done is to take a small bit of information that we don't really understand and run with it. All the "eco people" and "antieco people" have turned into Chicken Littles who say, "The sky is falling." Which one of their skies is falling? Do we dare look up? Ultimately, the decision about whether to recycle or make one's own household chemicals is one that generally should be left to individuals and businesses. We can't legislate common sense or spend away a problem. Recycling is just a choice that should not to be made out of guilt or fear. Do it because you want to. The nice thing about recycling that may surprise you is that no matter how small or how little of an effort you make to be a little greener, it just makes you feel better.

There's no wrong shade of green. It's like sitting at a red light and not seeing that it has changed to green. Your passenger notices that it has changed and says to you jokingly, "Is that not the right shade of green for you?" You smile and move on through the intersection because when it comes to green, no matter the shade: *It's all good to go.*

CHAPTER 2

Recycling the Good, the Bad, and the Ugly

RECYCLING IS REPROCESSING used materials into new products or to prevent waste of potentially useful materials, thus reducing the consumption of natural resources and raw materials, saving energy, and reducing pollution. Recycling is the opposite mind-set of the "just bury it and forget about it" method we used most commonly in the past for waste disposal. Recyclable materials include many kinds of papers, plastics, glass, metals, and other solid items that we use around our homes and work. Materials to be recycled are either brought to a collection center or picked up from the curbside, then sorted, cleaned, and reprocessed into new products. Recycling can produce a constant supply of materials we use to manufacture products or to replace nonrecyclable resources with ones that are. For example, used paper often goes back into new paper, and plastic milk jugs can be made into plastic lumber, thus replacing treated wood used for outdoor applications such as decking for our homes.

Collection

There are a number of different systems and methods to collect recyclables from the solid-waste stream. These methods are a tradeoff among public convenience, ease of use, and cost. The three main systems of collection are drop-off, buy-back, and curbside.

Drop-Off

Drop-off centers require that the recyclables be brought to a central location to be recycled. Drop-offs are the most commonly used collection system, but often they have a higher cost and energy usage associated with them. It does seem silly to make a special trip, or spend my time and energy hauling a few cents' worth of recyclables across town, that's costing me over a dollar just in gas. It's important to drop off your recyclables when you have more errands to do than just recycling.

Buy-Back

Buy-back centers purchase recyclable materials. This works well with high-end recyclables such as aluminum cans, copper, and most other metals. Unfortunately, government subsidies are often necessary to make buy-backs of low-end recyclables viable. More often than not, many recyclables wouldn't have a market if it weren't for tax credits, rebates, and incentives. If your business needs those breaks to be in the market, then the risk may be too high. The truth about tax credits and incentives is that we the taxpayers and consumers ultimately are paying for these incentives, so the costs haven't changed, just the distribution of debt and cost.

Curbside Collection

Curbside collection basically is waste collection at the curb that includes collection of commingled materials and/or source-separated recyclables. Mixed solid waste contains both recyclables and waste materials mixed together (commingled). Commingled collection materials go to a materials recycling facility (MRF) where the desired materials are sorted out and cleaned. Source separation is the other extreme, where each material is cleaned and sorted prior to collection. Once commingled recyclables are collected and delivered, the different types of materials must be sorted out. The sorting is done in a series of stages, some of which may involve automated processes and others of which may involve hand picking, which costs a lot more.

In automated recycling plants, machinery separates the recyclables by weight (e.g., lighter paper and plastic from heavier glass and metal). Cardboard is removed from the mixed paper, and all plastics are collected together. Strong magnets are used to pull out ferrous metals such as steel and iron. Finally, glass must be sorted by hand based on its color (i.e., brown, amber, green, blue, and clear).

Which Way Did It Go?

Plastics

With a little bit of care, most plastics can be recycled, and collection of plastics for recycling is ongoing. Plastic recycling does have one main problem (in most cases), and that is that plastic types must not be mixed for recycling. However, it is almost impossible to tell one type of plastic from another. The plastic industry has responded to this problem by developing a series of identity markings placed on the bottoms of plastic containers. Virtually every container made of plastic now should be marked with a code. This code must be molded into the plastic and located on the bottom surface of the container. Ideally, the entire container should be made of the same plastic to avoid confusion, but often the caps are of a different type. Caps should be marked separately, but few are, so the best thing to do is to put them aside. Plastic grocery and produce bags often are collected in barrels at grocery stores and usually end up as plastic lumber. Collection is not particularly profitable. Many products, such as compact discs, video tapes, and computer discs, are made from mixed materials that can't be recycled unless they are first disassembled.

Glass, Steel, and Aluminum Cans and Foil

Glass, steel, and aluminum are easy to recognize and recycle. Glass bottles must not be mixed with other types of glass, such as windows, light bulbs, mirrors, glass tableware, cooking glass, or automobile glass. Ceramics such as coffee mugs contaminate glass and are difficult to sort out. Clear glass is the most valuable. Mixed-color glass is nearly worthless, and broken glass is hard to sort.

It is no longer necessary to remove labels for recycling. To save water, clean only enough to prevent odors. Unlike with plastics, the high temperature of glass and metal processing deals easily with contamination.

Scrap aluminum is accepted in many places, and often you can get reimbursed for your recycling efforts.

Aseptic Packaging

The square boxes used for liquids are called *aseptic packaging*. Such packages are made from complex layers of plastic, metal, and paper. The actual recycling process is very expensive and therefore is available in only a very few places.

Because of the difficulties, only an insignificant fraction of aseptic packages are currently recycled.

Paper

Most types of paper can be recycled. Newspapers have been recycled profitably for more than a half century, and recycling of other paper is commonplace. The key to recycling paper in an affordable manner is to collect large quantities of clean, well-sorted, uncontaminated and dry paper.

It is important to know what you are buying in a paper product, and for this reason, virtually all paper products should be marked with the percentage and type of recycled content. For example, *postconsumer* paper is paper that is made from what you and I return to recycling centers. From a recycling point of view, the more postconsumer paper the better. Soybean-based inks are gaining favor these days as a renewable alternative to toxic inks.

White Office Paper

One of the highest grades of recyclable paper is white office paper. Acceptable are clean white sheets from the likes of laser printers and copy machines. Staples are okay. White office paper that is wet or otherwise damaged may be downgraded and recycled with mixed and/or colored paper.

Corrugated Cardboard

Cardboard is easy to recycle unless it is contaminated, such as greasy pizza boxes, which are not acceptable. In some areas, cardboard must be free of tape, but staples generally are okay.

Newspaper

Newspaper is widely available and of uniform consistency, which makes it valuable. The entire newspaper, including inserts, is acceptable for recycling, except for such things as plastic, product samples, and rubber bands. Newspapers can be stuffed in large brown grocery sacks or tied with natural-fiber twine to ease their handling. Other brown paper bags and craft paper may be mixed with newsprint.

Phone Books

Printed in your phone book should be information on the source and type of paper used, the nature of the binding, and where phone books can be recycled locally.

Mixed Paper

Mixed paper is a catch-all for types of paper not specifically mentioned above. Everything you can imagine from magazines to packaging is acceptable. The paper still must be clean, dry, and free of food, most plastics, wax, and other contamination.

Paper That Is Hard to Recycle

Paper that can't be recycled as normal mixed paper includes food-contaminated paper, waxed paper, waxed cardboard milk and juice containers, oil-soaked paper, carbon paper, sanitary products or tissues, thermal fax paper, stickers, and plastic-laminated paper such as fast-food wrappers, juice boxes, and pet food bags.

Appliances (Refrigerators, Washers, and Dryers)

Most older refrigeration equipment contains a refrigerant called *Freon*, which is a chlorinated fluorocarbon (CFC) that facilitates refrigeration. If you are throwing away an old refrigerator, heat pump, or air conditioner, please be sure the CFCs are drained out and recycled first. Use only a hauler who will perform this important service. Call and ask about this before you let anyone take your old equipment away. In most areas, there are eager people who will pick up your appliances for parts and/or recycling at little or no cost to you and will deal with the CFCs properly.

Single-Use Batteries

Often we just tossed old batteries in the trash, but increasingly, these batteries are now being collected, although not actually recycled. Often they are simply put in a more expensive landfill. Use rechargeable batteries as often as you can, and invest in "smart" chargers that are best for battery life.

Rechargeable Home Batteries

Rechargeable batteries are used commonly in portable telephones, computers, power tools, shavers, electric toothbrushes, radios, videotape recorders, and other consumer products. There are a variety of different battery types, some of which contain quite toxic materials.

Nickel-cadmium (Ni-Cd), nickel–metal hydride (Ni-MH), lithium ion (Li-ion), and small sealed lead (Pb) batteries all can be recycled. Several states now prohibit consumers from dumping rechargeable batteries into the normal trash. Stores such as Radio Shack either will take your batteries to recycle them or will help you to find a local source that does.

Motor Oil, Tires, and Car Batteries

All three of these products are big environmental problems, but all three are easily recycled. Used motor oil contains heavy metals and other toxic substances and is considered hazardous waste. Recycling used motor oil is easy, though. Typically, you place the used oil into a plastic milk jug and clearly mark it "Used motor oil." Many quick-lube shops will take used oil (the industry association encourages it). Some states have laws requiring any business that sells oil to take used oil back from consumers.

Normally, you must pay a fee to dispose of a tire. Improperly disposed of tires hold water and thus breed mosquitoes and transmit disease. In addition, large caches of old tires have been known to burn when not protected properly, with huge negative consequences for air quality.

Your old car battery may be worth money. Even if not, any car parts shop will take it.

Computer Printer Cartridges

Most printer cartridges are easily recycled, refilled, or rebuilt. Printer vendors sell their printers cheaply and make their money from selling the cartridges and other supplies. Before you buy a printer, make sure that the cartridge is refillable.

Buried Treasure in Our Trash

In the near future, I see modern landfills as a great way to bank or store less marketable resources for future use, lock away carbon, or add value to the carbon by turning it into energy. Today's landfill have greatly reduced the potential for risk once associated with older landfills. Today's landfills are built on a foundation of several feet of dense clay covered with a thick geotextile liner that is air- and watertight. This textile layer is covered with several feet of gravel or sand. As the waste is spread out at the end of each day, layers of dirt or other inert materials are used to cap it. All landfills produce leachate (liquid waste), which comes from the decomposition of waste within the landfill and water from the natural environment. This liquid must be dealt with. Modern landfills minimize fluid going in, such as rain, by covering areas that are not currently operational. All leachate that is drained out is sent to wastewater plants for treatment and purification. In high-tech and experimental landfills, the leachate is circulated in and out of the landfill to encourage faster decomposition of the organic waste. In the future, as resources buried in landfills find value, they will be mined, and the land on which the landfill was located will be reclaimed. Landfill mining, reclamation, and excavation are happening on a small scale today in Japan and Europe, where land values are high.

Banking on Trash

When it comes to banking or storing carbon, modern landfills can do a great job. As things break down, decompose, or decay into their basic elements, the mass of those items is changed into gases and whatever. When this happens in the open air, where oxygen is present, the process of decay is more likely to produce such gases as carbon dioxide (CO_2). The mulch we use in our flower beds and gardens is made from wood breaking down into CO_2. Even the kitchen waste, grass clippings, and paper we use to make compost in an aerobic (oxygenated) pile produce CO_2. With plastics, it's the sun's ultraviolet radiation that helps to speed up decomposition and again produce CO_2. Many metals fall prey to oxidation, and oxidation is caused in part by water and the oxygen found in the air we breath that helps to create an electrochemical reaction that eats away at the metals. The process is even more aggressive if a salt or acid is put into play, making the electrochemical reaction even more aggressive. Metals that are high in iron content oxidize (rust) very easily when unprotected, whereas metals such as aluminum, brass, copper, nickel, tin, and even gold hold up well in these conditions because they are less reactive to oxidation or even form their own protective coatings called *oxides*. In a covered landfill away from air (oxygen) and sunlight, metals and plastics can take up to hundreds of years to decompose. The plastics and metals in today's landfills have little value because they are mixed with low-value trash that we actually had to pay someone to take and bury out of site and out of mind. To recycle those plastics and metals now would take a great amount of energy and money, so unless their value skyrockets, don't expect them to be mined out any time soon. In the short term, therefore, we know that there is a resource there that might be reused in the future. The things that decompose rather quickly in terms of the life span of a landfill are the organic compounds called *biomass*. Biomass is made of all kinds living or deceased organic material. As a waste source, biomass refers to plants and animals that are raised and harvested for our use and the waste produced from them in liquid or solid form. As biomass waste decomposes or decays into its most basic elements, the mass of the item is changed into gases and simpler, less complex molecular structures. Above-ground decomposition of biomass produces CO_2 gas via its interaction with oxygen and things that feed on it such as insects, some animals, fungi, plants, and aerobic microbes.

Aerobic Microbes

Aerobic microbial interaction with biomass involves the same process as when yeasts consume the starches and sugars found in biomass and in so doing produce CO_2 and ethanol (alcohol). In a compost pile, aerobic bacteria do their job by converting carbon to CO_2, nitrogen to nitrates, and ammonia to ammonium, all of which help to make good rich soil.

Anaerobic Microbes

Anaerobic microbes are some of the oldest forms of life on earth and are responsible for the breakdown of organic material in the absence of oxygen. They produce a biogas as a waste product. Anaerobic decomposition occurs naturally in swamps, water-logged soils such as in rice fields, deep bodies of water, and the digestive systems of termites, large animals, and even humans. Anaerobic processes can be managed in an airtight environment such as a landfill or a covered lagoon used to store manure for waste treatment.

The anaerobic biogases (mostly methane) are called *landfill gases* (LFGs), and they are created by the anaerobic microbial feast that's occurring on the decaying biomass waste buried within the landfill. The use of landfill gas for energy also has the added benefit of offsetting the use of fossil fuels such as coal and natural gas. Landfill gas emitted from decomposing garbage is a reliable and renewable fuel option that remains largely untapped at many landfills across the United States despite its many benefits. Generating energy from landfill gas creates a number of environmental benefits.

Landfill Gas to Energy Is Good for Many Reasons

- It helps to keep methane from escaping into the air. Methane is a potent heat-trapping gas that is much worse than the greenhouse gas CO_2. Using it to produce electricity is a great way to further offset the use of nonrenewable resources such as coal, natural gas, and oil.
- There are many cost-effective options for reducing methane emissions while generating energy. Methane gas can be used to make another alcohol-based fuel called *biomethanol*, which is made synthetically from the methane gas that's naturally occurring from methane-producing microbes that thrive in oxygen-free environments such as landfills and covered sewage lagoons.
- Projects help to reduce local air pollution.
- Projects create jobs, revenues, and cost savings.
- Landfill gas from 1,000 landfills could power nearly a million homes.

What a Gas

Municipal solid-waste landfills are the second largest human-generated source of methane emissions in the United States. Given that all landfills generate methane, it makes sense to use the gas for energy generation rather than flaring it off to convert it into CO_2 while wasting its heat energy or emitting it as a poisonous gas to the atmosphere. Methane is a very potent greenhouse gas that is over 21 times stronger than CO_2. Methane also has a short atmospheric life and causes all kinds of undesirable chemical reactions as it breaks down. Turning methane emissions from landfills into energy is one of the best ways to achieve a near-term beneficial impact in mitigating energy cost and effects of methane emissions. When methane is used as a fuel and combusted, it's converted to water and the much less potent carbon dioxide gas (which is 21 times less damaging to our air). If we use the methane from a landfill to make energy, those emissions are not considered to contribute to global climate change (they are carbon-neutral) because the carbon is being emitted from recently living biomass. Not all biomass can be considered a biofuel; coal, oil, and natural gas are fossilized biomass fuels. Their origin is from ancient biomass that has been locked away from our current carbon cycle for millions of years. The combustion of ancient biomass may disturb the current carbon balance in our atmosphere, and it's a known fact that every action has an equal and opposite reaction. Today, our governments have banned the disposal of green waste, that is, grass clippings, leaves, limbs, and branches, in our landfills under the guise of saving landfill space. Recycling our green waste into compost and mulch produces the same amount of carbon dioxide emitted as a result of the natural decomposition of the organic waste materials outside the landfill environment. Thus, having that waste in our landfills again may help us to make even more energy and cause no more harm than if we just allowed it to rot on the ground. This is just something to think about.

Benefits to the Local Economy

Landfills can generate revenue from the sale of the gas or energy they produce. Use of landfill gas also can create jobs associated with the design, construction, and operation of energy-recovery systems. Landfill gas projects involve all kinds of outside labor, and much of the money generated is spent locally for drilling, piping, construction, and personnel. The economic benefits from locally produced energy include employment, local sales, and tax revenue. An investment in locally produced energy helps manufacturers to use landfill gas as a direct replacement for more expensive fossil fuels such as natural gas. Some companies will save millions of dollars by investing in landfill gas energy.

How Landfill Gas Works

Landfill gas to energy involves collecting the combustible landfill gas. The gas, in turn, releases the heat energy that is stored in its chemical bonds through combustion. During combustion, the organic compounds in the landfill gas react chemically with oxygen and pressure to produce heat. This process releases the heat energy stored in the chemical bonds by breaking those bonds apart to form water vapor, carbon dioxide, and other less volatile compounds. When we ignite landfill gas in a reciprocating engine, gas turbine, or boiler to generate energy, we also reduce pollution associated with the use of the fossil fuels that normally would've been used to produce that same amount of energy.

Landfill gas is extracted from landfills using a series of wells and a blower or vacuum system. This system moves the collected gas to a central point, where it can be processed and treated depending on the ultimate use for the gas. From this point, the gas can be used to generate electricity or replace fossil fuels in industrial and manufacturing operations, or it can be upgraded to pipeline-quality natural gas.

Recycling Half-Truths

There's One in Every Crowd: Recycling Is Not Cost-Effective

Critics of recycling dispute the benefits of recycling because of its cost and suggest that proponents of recycling often make matters worse and suffer from a lack of proof that recycling works or makes economic sense. Recycling critics most often argue that the cost and energy used in the collection and transportation of recyclables don't add up when we compare them with the cost and energy saved in the production of nonrecycled materials. For example, paper recycled back into paper via paper pulp can be recycled only a few times before the recycled material degrades and prevents further recycling. Thus recyclables become less recyclable when we recycle them. Furthermore, most recyclables would not have a market if it weren't for tax credits, rebates, and other incentives. If the recycling business needs these breaks to be viable in the market, then the risk may be too high. The truth about tax credits and incentives is that we the taxpayers and the consumers are ultimately paying for those breaks, so the costs haven't changed, just the distribution of debt and cost. Proponents of recycling dispute these claims, and the arguments from both sides have led to enduring controversy.

Recycle Because We're Out of Landfill Space

The waste-disposal industry has moved to using larger landfills, partly because of Environmental Protection Agency (EPA) regulations and the cost-effectiveness of consolidations and mergers. At the same time, the number of operating landfills has fallen sharply, so we now have to move our trash greater distances than ever before. Landfill space actually has increased and trash has become an interstate business because we now haul it to new high-tech mega-landfills. Our focus has been on the number of landfills rather than on their capacity, so it seems that we may be running out of space. When we see a barge filled with garbage out at sea or trash trucks driving down our interstates, it has less to do with not having a place for it to go and more to do with the inability or lack of desire on the part of the disposing entity to pay for disposal. Our landfill capacities have increased, but so have our costs of hauling the waste over such long distances. The truth of the matter has more to do with economics than with how much space is available in our landfills. As landfills raise their disposal fees (to over $70 a ton in some areas), recycling makes even more economic sense and is becoming profitable. Therefore, regardless of landfill space and cost, there are very real economic incentives to reduce by recycling what goes into our landfills. Recycling is essential to everyday life because the *cost* of landfill space has increased regardless of capacity. Recycling offers real economic savings based on disposal costs alone.

Recycling Uses More Energy and Creates More Pollution Than Using Virgin Resources

When local manufacturers use local recyclables, this is a win-win situation for the local economy. Now, in order to develop a market for recyclables, as well as the products produced from those recyclables, we first need to find ways to use recyclables and finished goods in the local economy that has produced them. Home owners first have to generate the recyclables, and then they have to be collected and partially sorted. Next, the processor purchases or sometimes gets the recyclables for free from the collection facility; therefore, the collection facility gets a return on its investment or an assurance that the collected goods are being truly recycled. The processor adds value to the recyclables by further sorting them, removing the nonrecyclables, and putting them into a marketable state (i.e., baled, crushed, or ground) for easy handling, shipping, or storing. The processor now can sell the processed recyclables to local manufacturers to get a return on its investment in processing them. The manufacturer then sells its finished products to retailers, who, in turn, sell those products to the same home owners who originally generated the recyclables. Finished products sold outside the local economy bring further economic benefits as well as energy savings. The finished goods made from recyclables often cost less to ship than raw recyclables because of their overall density. This all sounds like a simple supply-and-demand model, but there is a twist within our model. The twist is that when all the parties are working together in keeping the energy and money in the local economy, that stream of money and energy make those areas of the local economy more sustainable. To make the local system even more sustainable, the same manufacturers could sell their goods at a wholesale price to local home owners, thus rewarding them and the local community for their recycling efforts. Now the manufacturer has an even more loyal and steady supply of local recyclables for its operation. In turn, the home owners get lower-cost goods to use in their homes, thus encouraging them to be more vigilant in their recycling efforts. Thus, if recycling ever uses more energy or creates more pollution than virgin resources, it has more to do with the mismanagement of what's being recycled.

Wants versus Needs

Ultimately, when we talk about recycling, we are talking about balance. Something that is very important to every living thing that shares our planet, balance is the very rhythm of life. The things we leave behind and the things that we do affect that balance. All living things depend on that fine line of a sustainable balance for survival. As humans, we are stewards of the earth, and the symphony of life is directed by our fingertips and every move. Each breath, each action, and even each thought we have affects our perception, personal environment, and the external environment surrounding us. Humankind is a force of nature. Modern success has relied completely on our ability to make choices, and what we choose now will affect our success in the future. It shouldn't be all about politics or policies. It's all about needs, basic economics, and the conservation of natural resources. It's all about how we use our resources and what types of resources we use. Wants or needs alone did not cause this problem; it was much more the size and frequency of those wants or needs. The use of recycled resources and conservation of natural resources ultimately provide us with another shade of green where our needs do outweigh our wants.

Recycling and the Poor

Throughout cities in Asia, Africa, and Latin America, varying numbers of poor individuals survive by salvaging materials from the waste stream. These people recover materials to sell for reuse or recycling, as well as diverse items or their own consumption. Scavenging is a widespread phenomenon; you can find some level of it on any street in any city, probably even in your own hometown. Regardless of where it takes place, scavenging for survival in garbage dumps of the third world is a problem of its own. The world's poor faces multiple hazards and problems because of their daily contact with garbage. Scavengers are usually associated with dirt and disease and are perceived as a nuisance, a modern-day leper. Imagine in this modern age being thought of as less than human or even as criminal. Scavengers survive in a hostile physical and social environment. The poverty found in most developing countries forces the poor to make the most of the resources available to them. The poor have creative ways to satisfy their own needs, including the recovery of items not necessarily part of the waste stream. Fixing the problems that cause such severe poverty around the world has to start somewhere, and when even the smallest of economies is itself more sustainable, it is more plausible and possible to have a positive effect on the rest of the world. Poverty exposes the most basic of human needs, namely, the need for survival, and the best argument for recycling therefore may come from our poor because when push comes to shove and society's safety nets fail, *one person's trash becomes another person's treasure*.

When Recycling Goes Down a Questionable Path

Recycling makes good sense when the resources and the markets in which it precipitates are close by, thus keeping those natural resources near their point of use and keeping money in the local economy. On the other hand, when we ship recyclable goods overseas and then buy them back in finished goods, we are in many ways hurting and shrinking our own economy, as well as risking even more environmental harm than if we just stuffed them in a landfill locally. In my opinion, a good example of recycling gone wrong would involve my local recycling (collection) company shipping hundreds of tons of cardboard boxes to a third-world country. Now, the manufacturer in the third-world country who bought the cardboard sees this as a good investment because local fiber resources such as cardboard are hard, if not impossible, to find. The shortage of fiber resources in the third-world country is due in part to its poor management ot its own natural resources and because the recyclables it gets from us are so affordable (in part because our tax dollars subsidized those costs), so there is little incentive for that country to manage its own natural resources in a sustainable manner. The local rural population is more concerned with burning their trees (a natural fiber resource) to clear land on which to raise cattle to feed the workers who manufacture goods using the recycled cardboard that came from our trees. Therefore, basically, we're recycling here to save trees, whereas in fact our cheap cardboard in small way may be encouraging the destruction of another few acres of forest elsewhere. Each time we send our recyclable materials away, we risk doing more harm, no matter how good our intentions.

So how can the third world compete with our local industries and still afford to ship goods around the world? Well, labor in the third world is cheaper, and regulations and safety concerns are nearly nonexistent. So once again it is all about economics and the fact that we the consumers demand inexpensive goods and lots of them. Take in account the slack regulations in other countries, and you begin to realize that we actually may be causing more environmental harm than if we had just once again tossed things into our own landfills. The world really is a small place, and as our planet turns each day, we cross paths in space with our neighbors thousands of miles away. We've all heard the phrase, "What goes around comes around," and that's exactly what's happening. Recycling and improving our own environment here may be more harmful to us because we are blind to how those materials are being processed overseas, and the very system of exporting recyclables offers no incentive for the nations that use them to manage their own natural resources properly. We take for granted so many things, and it's easy not to see a potential threat even when it is in front of our faces. *It's hard to see the forest when the trees are in the way.*

Re-recycled Investment

No matter what your reasons are for recycling or not, the potential value of the things you throw away can add to or take away from your own financial standing. Most of us don't feel that our waste affects our finances because it is such a small part of our personal budget. However, if we look at those costs over a lifetime and then multiply them by the millions of people who also create waste, and add in the taxes we pay to subsidize recycling efforts, the costs become astronomical. My point here is that however small or unimportant solid waste may be to us as individuals, it adds up to be an enormous burden for our modern society as a whole. Waste and recyclables both have a cost to us as consumers and taxpayers. We pay people to pick up our waste from our homes, businesses, and even along our highways (removing litter waste alone has an enormous cost). Recycling can cost us even more money because it takes a lot of our time to do, and we pay extra for the recycling service directly by fees or indirectly via taxes that are used to fund those recycling efforts. Waste is part of our lives, and it stays around for years and years if we bury it in landfills; aside from making it into landfill gas for energy generation or banking on it to have value in the future, our waste is a very high-risk investment. When we recycle, it's a low-risk investment partly because of the markets we are creating today, which are building an ever-growing demand for recyclables. As energy prices rise and natural resources become more expensive, our investment in recycling offers us a viable economic tool to hedge against inflation.

Recycling Your Waste Stream at Home

Imagine recycling the waste stream in your home or business. With a little time, tools, and know-how, you could take things that you once sent to someone else to recycle and turn them into products you can reuse yourself. You can save money by replacing things you might normally buy with items you made for almost free straight from your trash or recycling bin. Better yet, how about creating items from your recyclables and selling them to your friends and neighbors and making money from your own trash?

Cash in Your Trash

When we can recycle things close to the source (such as in our own homes) of those recyclables or waste, we use less energy, save time, and reduce the effect that outside forces have on our lives. Adding value to recyclables in our own homes is a very good thing, and gaining another source of income from them is even sweeter. When your own shade of green not only saves resources but also is a source of revenue—putting money in your pocket—all of a sudden your own shade of green has turned golden.

CHAPTER 3

Greening Up Household Cleaners

SOME OF US CLEAN our homes every day or once a week, and then you got the nearly never people. We all have different ideas on what clean is. Clean knows no difference between the haves and have-nots. I personally have been in homes that on the outside look like the owners had a picture-perfect life only to find them cluttered, dusty, and full of fleas. The owners seemed to be aware of this, but still never did anything to change their home environment and, in fact, seemed happy with it the way it was. I've also been in shacks that were so free of dust with everything so perfect and in its place that I would take off my shoes before entering. Our homes are our personal space. How often we clean them and what we clean them with are wholly our own personal choices.

What We Are Cleaning

Basically, when we clean our homes, we are removing foreign matter, substances, waste, and stains that we have either brought in or created by just everyday living. Sometimes, when we clean, it has more to do with not how things may look, but rather to keep our inside environment free of allergens, pollutants, and clutter. This is why we need to be sure that what we are cleaning with isn't worse for our indoor environment than what we are cleaning out of it. What good do we do if we use something that's more toxic than the muckety-muck we just removed?

How Do Cleaners Work?

Whether you're cleaning the dirt, dust, gunk, and stains off a chair you found at a flea market or the layer of sticky stuff off of whatever got spilled on that retro coffee table from the 1970s, you need to know what cleaners are and how they work if you're going to be successful with as little effort and damage as possible to whatever you're cleaning and to your wallet to boot. Choosing which way to remove what needs to be removed can be a difficult task in itself. The object of the method or madness you use is to separate the layer of dirt, dust, gunk, and stains from the surface or substrate of the object you want to clean. Housecleaning is work no matter what technique you use. When you dust, you can try to vacuum it away or wipe it away with something that makes dust stick to it, such as an oiled rag, lamb's wool, or a static-charged thingamajig. For other types of household cleaning, most of us have used those store-bought cleaners that offer a wide variety of compounds to clean surfaces and substrates. They are designed to degrade or break down the dirt, dust, gunk, and stains or to destroy (loosen) the adhesion of the coating or stain to the substrate. They must accomplish this without damaging or chemically interacting with the materials that make up the surface or substrate that is being cleaned. In addition, you want your household cleaners to be free of chemicals that are highly flammable and are low in volatile organic compounds (VOCs) or poisonous. Most cleaners and stain removers have a solvent effect in which small molecules found in the cleaner's ingredients easily penetrate into whatever you are trying to clean off and cause the chemical that bonds it to loosen or expand rapidly, pulling it away while removing it from the surface or substrate of the item being cleaned. You want what you just cleaned away to either stick to whatever you used to remove it or simply to just rinse away. The main ingredient in most cleaners is *surfactant*. Surfactants basically make water wetter. Yup, I said wetter water. We've all seen water bead up on hard surfaces naturally; this is a result of *surface tension*. Surface tension slows the wetting of the surface of what you are trying to clean and therefore makes the cleaning process all that much harder. A surfactant reduces surface tension so that the water can get what it touches wet while also lifting dirt, gunk, and stains from the surface that holds it to be carried away by a rag, mop, or just rinse water.

What's in Dirt?

Dirt is naturally occurring, unconsolidated, or loose material at the surface of the earth that is capable of supporting plant and animal life. In simple terms, dirt consists of three components: solids, liquids, and gases. The solid phase, which accounts for approximately 50 percent of the volume of a typical dirt, is a mix of *mineral* and *organic matter* and gives dirt its mass. Dirt particles fit loosely together (depending on particle size), leaving "open" spaces. The open spaces then are filled with water (making mud) or air. The water and air in dirt around our homes make up the other half of the dirt. All soils are made up of the three major components; however, the portions will vary. The mineral matter is derived from the weathering of hard rock at the earth's surface. Most dirt that we bring in our homes is made up of approximately 45 percent mineral content, and the number and sizes of mineral particles vary. In household dirt, the amount of organic matter (living and/or dead organisms) will range between 1 and 5 percent. Organic matter mostly consists of dead plant and animal remains. This decay in the upper layer of dirt, soil, or topsoil is the major source of nutrients for plants and other organisms. The higher the dirt's organic content, the higher is the quality of the dirt in terms of nutrients for plants and the retention of moisture. The downside to the home owner is that if you have a healthy yard with plenty of organic matter, it's going to make real nice mud that finds its way into your home, most often on your shoes. Just a small amount of mud can make a big mess of dirt. The largest part of the dust we have in our homes also comes from the same things that make up dirt, but, as we will see, dust can have a few more nasty things in it.

Even though we may not see dust around our homes very easily, it's there. We can find it simply by looking at the sun's rays as the sun shines in our windows. Using a duster sometimes can just spread or move the dust around.

What's in Dust Besides Dirt?

- Shed human skin cells (about half a pound per person per year in our entire lifetime)
- Paint particles (as paint wears, it becomes dust)
- Smoke from cooking and cigarette smoke (and its toxic by-products)
- Fabric fibers from your clothes, curtains, carpet, etc.
- Plant and insect parts
- Mold spores from inside and outside
- Shed pet skin, fir, and feathers
- Dust mites and their excrement
- Viruses
- Rodent and insect waste
- Pollen

- Bacteria
- Mineral fibers such as asbestos, fiberglass
- Organic fibers such as wood and paper
- Heavy metals such as lead, cadmium, and mercury

As you can see, the cleaning of the things we own has a greater purpose than just to satisfy our vanity, impress our guest and neighbors, or feed the compulsive disorders some of us share. What's in the very dust and dirt we have in our homes may threaten not only our health but also our very lives. People with weakened immune systems, asthma, and other illnesses can suffer unknowingly, get worse, and even die as a result of the silent and nearly invisible threats found in our dust.

What Makes Things Stick or Stain?

There are many ways in which things stain or stick to our possessions, and there are some slight differences in the mechanics of how they stain or stick, so I will just talk about how things stain or stick in general. The first aspect of being stained is that what's doing the staining must be able to creep or flow into the crevices of the object that is becoming stained or sticky. At a microscopic level, even surfaces that appear smooth may be very rough (imagine that surface as being full of mountains, valleys, and crevices). The staining stuff has to be able to seep into those crevices. To become a set-in stain, the material that's staining must be able to change its molecular structure from a fluid to a solid in some cases or leave behind enough pigment/coloration locked within the object's spaces. There are many ways in which this can be achieved. Reaction with oxygen in the air we breath, heating by sunlight, and thermosetting (hardening owing to heat) are all processes that can make a stain stick even worse.

Thus, what we can imagine, is that the staining material, made up of a fluid or very small solids, will seep or fall into the microscopic cracks and crevices of the object being stained. Then, through some outside mechanism (noted earlier), the stain hardens, cures, or sets in. The result is that the parts of the staining material that have bonded with the object at the microscopic level now act like hundreds of little anchors that hold onto where they landed. Once set in or hardened, those anchors are even more difficult to dislodged and remove. We consumers spend millions of dollars fixing/removing stains from our environments. In nature, stains are part of how the system works; they are a natural occurrence. In the human environment, stains are called accidents or mistakes. When stain prevention fails and we end up with a real bear of a stain, the next thing we need to do is to know what made the stain. Fighting stains is like fighting a war, and as with any war, knowing what you are fighting is half the battle. After all, you don't want to bring a knife to a gunfight. The simplest of stains can be made much worse or permanent if it is treated in the wrong way.

Can't You Smell That Smell?

The sense of smell is one of nature's most primal resources. We all need our sense of smell to survive. Our sense of smell can signal danger, induce hunger or attraction, or produce dread. A good example of a modern warning system that uses an odor and our sense of smell is under our very noses, as it were. We use our sense of smell to detect deadly gas leaks caused by natural and propane gas. The suppliers and marketers of those fuel gases add an odorant that is similar to the odor of rotting eggs or flesh. A gas leak in your home can be dangerous because it increases the risk of fire or explosion. Gas companies add the offensive warning odor so that it can be easily detected by most people. However, people who have a diminished sense of smell may not be able to rely on this odor safety mechanism. If you have a concern about your ability to smell the additive that signals a gas leak, you need to see a physician and use a different safety signal. A gas detector can be an important tool to help protect you and your family. We can detect spoiled food with odors that may help us to avoid being poisoned by bacteria. The smell of smoke helps us find fire so that we can control it or stop its spread. Some odors communicate less critical messages and can convey different results or meanings to different individual experiencing the odor and as well as how the person is receiving it. The aroma of a roasting turkey sends one message when someone is hungry and another very different one when that person has just finished a large meal. The smell of some soft cheeses such as Limburger may turn the stomachs of some people while making mouths water in others. What we smell and how we perceive it have a lot to do with how others around us react to an odor and what we're taught as children.

The Nose Knows

Our sense of smell is very complex. The brain can analyze hundreds of odor molecules. The average person can discriminate thousands of combinations of odor molecules. We breathe in airborne odors that travel to and combine with receptors in nasal cells. The cilia, hairlike receptors that extend from cells inside the nose, are covered with a thin mucous layer that dissolves or collects odor molecules in various forms. In the highest part of your nose, these molecules touch a group of nerves called the *olfactory nerves* that carry the odor messages to your brain. If your brain recognizes the odor, this means that you have smelled it before. If it is a new odor, your brain makes a record of it so that you will remember it the next time you smell it. The brain can store, recognize, and remember tens of thousands of different smells. It is not uncommon to smell something that brings back memories in vivid recall or to try to cover up an unpleasant odor with a more pleasant one we remember. Our noses and brain also can become numb or used to strong odors. This is often the case in people who work with animals or work at trash dumps. Our sense of smell is always adjusting to the environment we're in so that we can cope with and recognize other odors we may encounter, both pleasant and not so pleasant.

Raising a Stink

To remove an odor, you should use your nose to detect that odor. This means that you should concentrate on smelling or sniffing out the odor by using your sense of smell. Sniff around the area to determine where the smell is coming from because the molecules making the odor are going to be more numerous near the odor. Once you find the item or the area that is causing the odor, you should remove it and try to clean it or dispose of it. Wash the odorous area with warm water, and use an antibacterial soap unless the item is also stained. Remove stains first so as not to risk setting in the stain. Once the item is completely clean, then you can air out the area.

Smoke Odor Tips

Airing the area out requires the least amount of effort, so you definitely should open all your windows and doors for several hours to encourage the odor to lift from cushions, carpets, and other surfaces. If a particular piece of furniture reeks of smoke, bring it outdoors, place it somewhere dry, and let it sit for a few hours. Launder or dry clean whatever you can. This is the next simplest way to deal with smoky odors such as cigarette smell. You also can use a vacuum cleaner to suck the smoke molecules out of furniture and upholstery in your house or car. Sprinkle baking soda over the smoke-infused area, and let it sit for a few hours or overnight. Then use your vacuum cleaner to suck up the soda, smoke odor and all. Use liquid cleaners or a vinegar and water solution with a cotton rag to scrub smoky residues off linoleum floors, glass, and wood surfaces. The smooth surface and heat of light bulbs attract smoke, and each time you turn the light on afterward, the same heat releases odors from the smoke's residue. Clean light bulbs carefully with a diluted vinegar and water solution and a cotton rag.

Again, you'll be required to use your nose to sniff out the spot if you can't see it. Once you find the stain, you should blot the area with a white paper towel and then use white vinegar to completely clean the area, which will partially get rid of the odor. Once it is completely dry, you can cover the area with baking soda, let it sit a few hours, and vacuum away the residue.

Refrigerator Odor Tips

For odors in your refrigerator, the first thing that you have to do is clean out all the spoiled or odor-causing items. Meanwhile, save your good perishables in a cooler. Then you should clean all the trays and the inside of the refrigerator with a mixture of soap and water or baking soda and water. Rinse the refrigerator out completely with clean warm water, and wipe it dry. Clean all items that you want to keep with soapy water and then rinse and dry them with a clean rag. Replace everything, and leave an open box of baking soda in the clean refrigerator.

Garbage Disposal Odor Tips

For smelly garbage disposals, clean what you can without putting your hand in the disposal; then pour vinegar and then baking soda down the disposal and rinse with warm water. You should regularly grind up some lemon, orange, or grapefruit rinds to keep a cleaner smell.

Basic Cleaner Types

Abrasive Cleaners

Abrasive cleaners are like liquid-sandpaper cleaners. They physically scratch off dirt, stains, and tarnish with friction as you rub the surface. They are composed of either particles or physical abrasives such as sandpaper, steel wool, or pumice. The finer the particle, the less abrasiveness it has, whereas the coarser and denser the particle, the more abrasive it becomes. Baking soda and salt can be used as abrasives. Baking soda is finer, less abrasive. Salt is more abrasive. Abrasives can dull glossy surfaces and change both the look and texture of surfaces.

Mild Abrasive Cleaners

These materials include plastic mesh pads, nylon-coated sponges, and fine-metal "wools." Mild abrasives are used on pots and pans, oven interiors, and enamel sinks. Use as directed to remove stains from surfaces of furniture and countertops, and be aware that abrasives, even mild ones, will scratch fine, hard, smooth surfaces if you rub too hard.

Moderate Abrasive Cleaners

These cleaners should be used sparingly. They are made of such things as pumice blocks and fine steel wool. Steel wool is actually graded 0000 (superfine), 000 (extrafine), 00 (very fine), 0 (fine), 1 (medium), 2 (medium coarse), and 3 (coarse). Steel wools rated 00 and finer should be used lightly on pots and pans when needed to remove burned or encrusted foods and grease. They are often used on burned-on spills.

Strong Abrasive Cleaners

These materials include medium and coarse steel wool, metal mesh cloths, metal brushes, coarse pumice, minerals, and sand. Use them on barbecue grills and untreated oven racks for stubborn deposits when damage to surface is not important.

NOTE Strong abrasives quickly abrade hard surfaces, making them coarse and thus more susceptible to dirt and stains later.

Acid Cleaners

Acid cleaners can be used to remove tarnish, alkaline discoloration, corrosion, and hard-water deposits from many surfaces and may have a bleaching effect.

CAUTION Acids can injure eyes, skin, and fabrics. Some acids are highly active and can eat through metal and etch surfaces such as porcelain enamel.

Mild Acid Cleaners

These include lemon juice (or citric acid) and vinegar (or acetic acid). They help to remove hard-water deposits from shower doors, mild rust stains, soap film, and tarnish from brass and copper.

Strong Acid Cleaners

These include dilute hydrochloric (muriatic) acid, dilute sulfuric acid, and sodium bisulfate. These acids are used in some toilet bowl cleaners and etching compounds. They remove hard-water and iron deposits as well as organic matter.

CAUTION Never mix acids with any other cleaner. It's very dangerous and toxic as well. Follow label instructions exactly. Do not get on skin, in eyes, or on other materials.

Alkali Cleaners

Alkali or alkaline cleaners remove both suspended heavy soil and grease so that it can be rinsed away.

CAUTION Alkalies can damage skin and fabrics and corrode and darken aluminum. Most (except baking soda) are toxic if swallowed.

Mild Alkali Cleaners

The most common mild alkali is baking soda (sodium bicarbonate). Soak burned food from pans in a solution of 2 tablespoons of baking soda per quart of warm water. (For heavier or sticky spots, sprinkle baking soda on a damp sponge, rub, and rinse, or make a paste of baking soda.) Baking soda can be used to clean glass, tile, porcelain enamel, stainless steel, chrome, and fiberglass tubs and showers. It also removes coffee and tea stains from dishes.

Moderate Alkali Cleaners

The most common of these is ammonia, which can be used in a solution of 2 tablespoons per quart of warm water to clean windows, glass, ovens, range burners, and greasy surfaces. Moderately strong alkali cleaners such as borax can be used in a mild solution consisting of 1 to 2 tablespoons per gallon of warm water to wash sinks and painted walls. This solution also can be used in laundering to remove odors and retard bacterial growth. It also removes soot and smoke and greasy dirt.

CAUTION Moderately strong alkali cleaners can have irritating fumes. They also can soften paint, especially latex paint, if they are too strong. Always use them alone because combining ammonia with other cleaners may produce lethal gases.

CAUTION Always wear rubber gloves to protect your skin, and always rinse thoroughly.

Strong Alkali Cleaners

The most common of these is washing soda (sodium carbonate), and it can be used in dilute amounts to soak greasy burners, pans, and BBQ grates.

CAUTION **Sodium carbonate is highly toxic. Do not get it on your skin. It also will darken and corrode aluminum.**

Very Strong Alkali Cleaners

The most common of these is lye (sodium hydroxide), and it is used to clean grease from ovens and to open grease- and hair-clogged drains.

CAUTION **Lye is very caustic and toxic. It can cause serious burns to the eyes or skin. Follow label instructions exactly.**

Bleaching Agents

Bleaches can oxidize and remove stains from surfaces and fabrics.

CAUTION **Rinse bleached items thoroughly and promptly to prevent fading.**

Mild Bleaches

Mild bleaches such as dilute hydrogen peroxide can be used in solution with water to help lighten stains on surfaces such as plastic laminate.

TIP **Hydrogen peroxide can be bought in 3% to 35% concentrations. It is a very powerful oxidizer in higher concentrations. We all know that water is H_2O; hydrogen peroxide is H_2O_2, so it is basically water with an extra oxygen attached.**

CAUTION **Rinse items thoroughly, and follow label directions.**

Strong Bleaches

Strong bleaches such as chlorine bleach (sodium hydrochlorite) remove stains and disinfect toilet bowls, trash cans, and other surfaces.

CAUTION **Sodium hydrochlorite may bleach dark-colored surfaces dull, such as the porcelain enamel finishes of sinks or tubs. Never mix with ammonia or other cleaners because the fumes produced are toxic.**

TIP **Do a colorfastness test to decide if a fabric or surface can be bleached safely.**

Colorfastness Hint

Dissolve 1 teaspoon of Clorox 2 in 1 cup of hot water. Soak a hidden area of the garment in the solution for 1 minute. Rinse and let dry. The garment is safe for washing if color does not fade or bleed.

TIP **If a stain is not removed within 15 minutes when bleach is used, it cannot be removed by bleaching, and further bleaching will only weaken the fabric.**

Detergents

Detergents can be alkaline or neutral. Neutral detergents are pH 7, meaning that they are neither acidic nor alkaline. Read the can or bottle first to determine a detergent's level of alkalinity.

Mild Detergents

These include hand dishwashing liquid. Detergents have surfactants that dissolve or lift dirt and grease away. Use in a solution of warm to hot water to clean washable surfaces such as countertops, appliances, fixtures, and floors.

Moderate Detergents

Moderately strong detergents use both surfactants and builders, so they dissolve heavier soils and grease. Use the least amount of detergent that will do the job. Low-sudsing types are easier to rinse off.

TIP To boost a detergent's grease-cutting ability, add an alkali such as ammonia or borax (i.e., boron or boric acid).

TIP To boost a detergent's soil- and odor-cleaning ability, add vinegar to neutralize the odor and soften the dirt.

CAUTION Most powdered detergents contain washing soda as a builder, and some are very alkaline, which can damage surfaces and irritate skin.

Solvent Cleaners

Solvent cleaners are readily available and are used frequently to dissolve household dirt and grim. A solvent is often the go-between that a stain or soil goes into so that it can be wiped or rinsed away.

Water is a solvent because it will dissolve most kinds of soils except oils. Alkalies, acids, bleaches, and/or detergents are often added to water to chemically react and loosen dirt and grim so that they can be removed with the water.

CAUTION Surfaces can be damaged by water. For example, wood can warp, wood finishes can soften or discolor, fabrics can shrink, and water can result in the growth of mold and mildew. Some materials soften or disintegrate in water, such as paper and sheet rock. Be extra careful around electrical components, where shock hazards can exist.

Organic Solvents

These include acetone, denatured alcohol, methyl/ethyl esters (biodiesel), petroleum distillates such as kerosene, mineral spirits, naphtha, dry-cleaning fluid, and turpentine. They are used to remove greasy soil or stains that will not dissolve in water, in addition to being used for cleaning surfaces that normally would be damaged by water. Some organic solvents remove waxes, finishes, and oil-based paints. Many are used in polishes, waxes, spot removers, rug cleaners, degreasers, and automotive exterior cleaners.

CAUTION Most organic solvents are flammable or combustible. Read and follow all product directions completely.

Store-Bought Cleaners and Their Potential Dangers

Always read product labels, and look for these signal words:

Danger! **Warning!** or **Caution!**

These words warn us about hazards posed by the product. It's common sense that danger means that a product is extremely hazardous, poisonous, flammable, or corrosive. Warnings or cautions signal a product that is somewhat less hazardous. Products listing no signal words are usually the least hazardous.

A product is hazardous when it contains one or more of the following properties:

- Flammable/combustible
- Explosive/reactive
- Corrosive/caustic
- Toxic/poisonous

The Dangers of Hazardous Household Products

Potential Health Problems and Injuries

- Mixtures of some hazardous products can produce dangerous vapors or fires.
- Products containing acid or lye can burn the skin and eyes.
- Exposure to some pesticides, paints, and solvents can make you feel sick, causing weakness, confusion, dizziness, irritability, headaches, nausea, sweating, tremors, convulsions, and even death.
- Repeated exposures to some chemicals can cause cancer, birth defects, and death.

CAUTION **Poisonings happen every day, with children and pets becoming ill and/or dying from eating or drinking toxic products in our homes. Many poisonous products can look and even taste good (e.g., automotive antifreeze).**

Prevent poisonings by

- Keeping products in their original containers when possible. If a product does not have its original label, label it yourself if you are sure of the contents.
- Not mixing products together. Dangerous reactions can occur when some materials are mixed.
- Making sure that products are sealed properly to prevent leaks and spills.
- Keeping containers out of the reach of children and pets.

TIP **Keep emergency numbers on hand as well as the number for the poison control center for your area.**

Ingredients to Be Aware Of

Nonylphenol Ethoxylates (NPEs)

When they're released into the environment, these chemicals can break down into toxic substances that can act as hormone disruptors, potentially threatening the reproductive capacity of fish, birds, and mammals. NPE is found in many cleaning products, especially detergents, stain removers, citrus cleaners, and disinfectants.

Antibacterials

Some antibacterial ingredients may cause skin and eye irritation, especially certain types with ingredients such as triclosan. Keep in mind that many products that carry the "antibacterial" label are actually disinfectants.

Ammonia

Ammonia is poisonous when swallowed and extremely irritating to respiratory passages when inhaled. It can burn skin on contact. It is frequently found in floor, bathroom, tile, and glass cleaners.

Never mix ammonia-containing products with chlorine bleach. This produces poisonous chlorine gas.

Butyl Cellosolve

This is poisonous when swallowed and is a lung tissue irritant. It is found in glass cleaners and all-purpose cleaners. (It is also known as butyl glycol, ethylene glycol, and mono butyl.)

Chlorine Bleach

Chlorine bleach is extremely irritating to the lungs and eyes. It is sold by itself and is found in a variety of household cleaners. (It is also known as sodium hypochlorite.)

Never mix chlorine bleach products with ammonia. This produces poisonous chlorine gas.

D-Limonene

This substance can irritate the skin. It is found in air fresheners.

Diethanolamine (DEA) and Triethanolamine (TEA)

When combined with nitrosomes, these ingredients may produce carcinogenic compounds that can penetrate the skin. They are frequently found in sudsing products, including detergents and cleaners.

Disinfectants

This is a widely used term for a variety of active ingredients, including chlorine bleach, some organic compounds, pine oil, and ethyl alcohol. They are regulated by the Environmental Protection Agency (EPA) as pesticides, and all have some health effects. Disinfectants are found in a variety of household cleaners. In addition, many products that carry the "antibacterial" label are also disinfectants.

Fragrances

Certain fragrances, especially synthetic types, may cause watery eyes and respiratory tract irritation. They are found in a variety of cleaners and air fresheners.

Hydrochloric Acid

Hydrochloric acid can severely burn skin and cause major irritations of the eyes and respiratory tract. It is commonly found in toilet bowl cleaners, masonry cleaners, and pool chemicals. (It is also known as muriatic acid.)

Naptha

Naptha can cause headaches, nausea, and central nervous system symptoms with overexposure. It is found in furniture and floor polishes and glass cleaners.

Petroleum-Based

Many ingredients are derived from petroleum, including some of the preceding, such as naptha, and they are commonly found in cleaning products and solvents.

Phosphates

Phosphates can escape sewage-treatment plants and septic systems to reach waterways and contribute to the overgrowth of algae and aquatic weeds, which can kill off fish populations and other aquatic life. Phosphates are found in automatic dishwasher detergents and some laundry detergents, where allowed.

Sodium Hydroxide

Sodium hydroxide is corrosive and extremely irritating to eyes, nose, and throat and can burn those tissues on contact. It is found in drain and oven cleaners and is used for making soap. (It is also known as lye.)

Sulfuric Acid

Sulfuric acid can severely damage eyes, lungs, and skin. It is freqently found in drain cleaners.

Still Not Sure?

If you're still concerned about the ingredients in a product, call the source. The manufacturer's name and address are listed by law on all cleaning products so that consumers can contact them with questions, comments, or problems. You also can request a Material Safety Data Sheet (MSDS) on the product that contains more detailed information on the ingredients and formulations used.

Hidden Dangers of Cleaners

There is evidence that links even light chemical exposure to the rising levels of chronic health problems in the general population. While most of us have known for a long time that high levels of exposure to some chemicals can be associated with certain chronic diseases, we now know that even minute traces of some chemicals can affect processes such as gene activation, hormone production, and brain development in our most valuable of resources—newborns.

It's a Bird, It's a Plane—No, It's a Superbug

To add insult to injury, more and more of us have become increasingly concerned about exposure to the germs that cause illnesses such as staph infections, bacterial meningitis, and strep infections, to name a few. Successful product marketing tells us of how new and improved products will protect our family's health by killing germs or viruses found on every surface we may touch. Antibacterial agents are doing the job on 99 percent of the germs. They wreak havoc on certain bacteria, but they don't touch viruses, and viruses account for a very large proportion of illnesses. The various products also are limited in the time they actually are effective. It hasn't been proven exactly how long they will work, but as soon as your clean hands touch anything else, you aren't protected anymore. The antibacterial soap you used isn't killing all germs your hands are coming in contact with after the fact. Dish detergents packed with antibacterial properties can't prevent the growth of bacteria on dishes but do protect your hands.

While we are definitely having some success in fighting bacteria, there are concerns about long-term problems with the use of these agents. Just as our overuse of oral antibiotics is creating bacteria that are resistant to antibiotics, researchers are worried that we are seeing the development of "superbacteria" that are resistant to the strongest of antibiotics. These "supergerms" are immune to our normal disease-fighting efforts. Another problem is that by trying to create a germ-free environment in our homes, we may be reducing our body's ability to fight infections. We all know the phrase, "What doesn't kill me makes me stronger." Well, this is the world we live in, and our bodies have learned to fight off major illnesses by fighting off minor illnesses. By far the most effective tool in reducing the spread of germs by both bacteria and viruses is proper hand-washing. That's right, just good old-fashioned washing of your hands with soap and water for 10 to 15 seconds can be your best defense against superbugs.

To Air Is Human

Most of us think that outdoor air is often more polluted than indoor air, but in reality, more often than not the latter is more polluted. In fact, our cleaning products play a role in generating some of that indoor pollution. The volatile nature of cleaning products and other household chemicals causes them to evaporate into the air we breathe every time we use them. This problem is even more apparent when we are cleaning in small spaces without fresh air, such as a windowless bathroom. In such situations, levels of chemicals in the air can become highly concentrated. The same chemicals we clean with also can end up in our household dust, making it a toxic cocktail, especially for people with asthma.

It's a Wash

Cleaning products and household chemicals are often rinsed down the drain and have an impact on our water quality. Those chemicals eventually are released with the effluent of sewage-treatment plants or from rural septic tanks into our ground and surface waters. We know that cleaning products that contain phosphorus or nitrogen can contribute to nutrient loading in open waters, leading to algae blooms that eventually tip nature's balance and destroy natural water quality. Most household cleaning chemicals survive the sewage system intact and are released into streams because sewage plants are engineered only to convert the biomass or organic waste to benign compounds. These chemicals often found in wastewater affect fish and wildlife exposed downstream of a sewage-treatment plant. The manufacturers even place warnings on their product labels about the near-term impacts of those chemicals. However, there are few or no warnings about the possible long-term connections to environmental or health concerns on those same labels.

A Little Big Deal

Children are disproportionately affected by everyday chemicals because of their smaller size and the fact that their organs and immune system have not fully developed yet. Chemicals sometimes can interfere with the development of neurologic and immune systems. Incidentally, babies and small children breathe more often and more deeply than adults, as well as consume more food and water per pound of body weight than adults, so their exposure to chemicals is concentrated. Children also play and crawl on the ground and frequently put their hands and objects in their mouths, thus ingesting chemicals from that contact directly into their bodies. Children are also at greater risk for accidental poisoning incidents associated with cleaning products because such products often look edible. The effects of some chemical exposures are not seen until the next generation, such as birth defects, low birth weight, or other harmful outcomes.

Take Control

Make Your Own Nontoxic Cleaning Products

Many recipes are available to make your own nontoxic cleaning products (see Chapter 8). Simple and inexpensive ingredients such as vinegar, baking soda, and borax can be used in many different ways for effective cleaning.

Use Less Toxic Products

It is not always easy to determine which cleaning products are less toxic. Simply having a "natural- or green-sounding" brand name does not exclude a product from containing chemicals that you may want to avoid. Clever marketing has "greenwashed" many things we buy at the store in an attempt to lead us to believe if it looks green it must be. *Buyer beware!*

The Easy Life

Most of our homes contain household products that we purchased to make our lives easier. Stores carry hundreds of brands of cleaners, detergents, polishes, paints, pest-control agents, and other products that promise to be fast, easy, and the best thing ever. We assume that a product is safe because it's for sale in a store in which we like to shop. Yet just about any household product can contain ingredients that can be harmful when we use them or dispose of them improperly. By understanding the products you use, how to handle them, and what alternatives you can make yourself, and knowing what else is available, you can make your home and environment a healthier place.

Your Shade of Clean

We don't want to just give up on all other types of cleaners that may or may not be completely green. Don't feel pressured to use all green cleaners because of a fad or misguided guilt. Many of us know how to use conventional cleaners safely and correctly and will continue to use them. When we use conventional cleaning agents properly we still can be very green. The point I'm trying to make is that no matter how we responsibly remove a stain, to save, reuse, add value to, and keep that item or thing in use, we saved it and our money from going into the waste stream. It's up to you and what shade of green you're comfortable with.

CHAPTER 4

Greening Up with Safer Pest Controls

Pests are part of life, and often they come in the form of an injurious or unwanted animal or plant. Pests can be tiny ants in our kitchen, weeds in our vegetable garden, or poison ivy on the side of our house. They can be merely annoying or a real health risk. At the same time, many of us are concerned that what we use to control pests can cause even worse problems than the pests themselves.

Weed-Free Lawns?

Do our lawns really need to be totally weed-free? Weeds bother some people more than others. Look at the lawn from the street. If its appearance is acceptable, do not waste your time going after those last few weeds. Excessive herbicide use intended to eliminate a few stubborn weeds usually does more harm than good. A 100 percent weed-free lawn is not a practical goal. Some things that look like weeds actually can help to keep your lawn green and healthy. And conversely, things that look like flowers and add color to your lawn could be hurting it.

Good Weed

Clover is a plant that most lawn owners immediately think of as a weed without considering the fact that it may be helping your lawn. Clovers actually fix nitrogen and add nutrients to your lawn. Clover even helps to keep real weeds away by competing with them. Therefore, when you see clover in your lawn, think about its benefits. People are becoming more concerned with natural resources such as water and its use and quality. In addition, clover lawns have been experiencing a revival. Around 50 years ago, a patch of white clover was considered the standard of excellence in lawns. Clover seed even was added to grass seed for people who wanted the best. There are two types of clover lawns: Pure clover is best for areas with low to moderate traffic, whereas mixed grass and clover work best for playing fields and high-traffic areas. Clover stays green all summer with little or no watering in most regions of the United States. Clover is drought-resistant, greens up early in spring, and remains green until the first frost. In warmer climates, it may remain green all winter. White clover grows just 2 to 8 inches tall and requires little or no mowing. You may prefer to mow in midsummer in order to deadhead old blooms or to prevent new blooming. Clover is a nitrogen-fixing legume, a plant that essentially creates its own fertilizer, as well as fertilizing other nearby plants. Grass that is intermixed with clover will be healthier and greener with less effort than grass planted alone. Most herbicides kill clover, but it outcompetes weeds, so you shouldn't need herbicides. Clover grows well in poor soils and tolerates a wide variety of soil conditions. Home owners who have been fighting clover as a weed get it for free if they decide to stop fighting and let it grow.

The Pest Things in Life Are Free

Pests come in many forms and are all living organisms that are a natural part of our environment. Pests include plants such as mold, mildew, and weeds, whereas those creepy pests such as insects, rodents, and bacteria top the list. Apartments and houses are often reluctant hosts to common pests such as cockroaches, fleas, termites, ants, mice, rats, mold, or mildew. On the outside of our homes we have weeds, hornworms (tomato worms), aphids, and grubs that can be a nuisance when they get into our lawn, flowers, yard, vegetable garden, or fruit and shade trees. Pests often become a hazard to our health, our family's health, and the health of our pets. We can choose from many different methods as we plan our strategy for controlling pests. For many pests, total elimination is very unlikely, but it is possible to control them. Knowing our options is what pest control is all about.

Pest Recognition

The first step is to identify the pest problem. This is the first and most important step in pest control—knowing exactly what you're up against is half the battle. Most pests are unmistakable, such as a cockroach, dandelion, or mouse. If you're not sure what kind of pest you have, ask a neighbor for help, look on the Internet, or go to the library—all can help you.

The Pest Offense Is a Good Defense

Controlling pests is far different from 100 percent pest elimination. Getting rid of all pests requires more invasive, repeated, and possibly hazardous treatments that are far from realistic. Moreover, it's much easier to prevent pests from finding their way into our lives than it is to control them. When we make the effort to prevent pests in the first place, the effect is more desirable and more effective. Pests are not our guests, so don't let them in.

Seal Cracks

The first defense is making sure that pests don't get into your house. Annually inspect the foundation and siding to find any holes or cracks that need to be sealed. Be sure to seal around exterior plumbing and electrical outlets. An adult mouse can enter your home through a hole no bigger than the diameter of a pencil.

Use Screens and Stripping

Window screens are excellent for keeping insects out of your house, even if you don't ever open the windows. Replace or repair any damaged screens on windows and doors. Check to make sure that windows and doors have good weatherstripping and are well sealed.

Control Lighting

Many insects are attracted to light, so just turn on exterior lights only when needed. Use yellow bulbs in exterior light fixtures because flies and moths are less attracted to yellow than to white bulbs.

Control Garbage and Clutter

Keep garbage in sturdy, tightly covered, and regularly washed containers. This prevents flies from breeding and reduces the attraction for insects. Avoid allowing clutter from clothes, papers, and other materials to accumulate in storage rooms because these provide good breeding grounds and hiding places for pests. Don't bring cardboard boxes into your home because they often carry cockroaches that are hitching a ride.

Eliminate Damp Areas

Critters and insects need water to survive, just as we do, so keep an eye out for wet or rotting wood in your house's structure, such as near the foundation. Moisture can attract certain insects such as termites and earwigs. Replace, paint, seal, or repair those wet spots.

What's Bugging You

Some insects are beneficial. In fact, many of them can be used to our benefit. Beneficial insects pollinate our flowers, and some can be used to reduce the populations of insects that we do consider to be pests.

Praying Mantis

Mantises are very deadly predators that capture and eat a wide variety of insects and other small prey. Camouflage coloration allows mantises to sit on twigs and stems while they wait. If you want to encourage mantises, you should use as few pesticides as possible and allow some vegetation to grow to provide cover for them.

Ladybugs

Ladybugs are the most common of all beneficial insects. Both adults and larvae feed on many different soft-bodied insects, but aphids are their main food source. Attract ladybugs by growing flowers and allowing a few dandelions to grow. To conserve ladybugs, use as few pesticides as possible.

Assassin Bugs

Assassin bugs are generalist predators that feed on a variety of insects. These predators are closely related to plant-sucking bugs. They have an elongated body with grasping forelegs and a pronounced head. Adult insects are brown in color and reach a length of 5 to 6.5 inches. Assassin bugs feed on many different insects, but because they have sucking mouthparts, they tend to feed on softer-bodied prey such as caterpillars. These bugs can be attracted to your garden by a cover crop such as a border of alfalfa. Assassin bugs have a higher tolerance to insecticides and pesticides than do most other beneficial insects.

Ground Beetles

Ground beetles eat soil-dwelling pests, including slugs and snails. Sometimes they also feed on plant pests. These beetles will seek cover in permanent pathways and perennial beds. The adult beetle ranges from ⅛ to 1¼ inches long. They are usually elongated, heavy bodied, and slightly or distinctly tapered at the head end. The ground beetle generally is dark, but it also can be purple, metallic green, or multicolored

Soldier Beetles

Soldier beetles are elongated, soft-bodied insects about ½ inch long. Their color varies from yellow to red with brown or black wings and trim. Most of the larvae are carnivorous, feeding on insects in the soil. Larvae stay in damp soil and debris or loose bark. The adults are also predators, eating caterpillars, eggs, aphids, and other soft-bodied insects.

Rove Beetles

Rove beetles are important predators of maggots and mites. Most rove beetles are slender with a very muscular and flexible abdomen. Adults range from less than 0.04 to 1.57 inches long (although almost all are less than 0.28 inch long). Adults are mostly nocturnal (meaning that they come out mostly at night). Providing a moist area, especially with decaying plant or animal material, can attract rove beetles. A good way to do this is to start a compost pile.

Flower Flies

Flower flies feed on aphids. Adults are ½ inch long. They resemble bees and are yellow and black striped or black and white striped. Adult flower flies can be attracted by pollen- and nectar-producing plants.

Lacewings

Lacewings are green and brown with large eyes relative to their head. Adults are generally ½ to ¾ inch long. These insects are generalist predators. They feed on aphids, mites, thrips, insects with soft scales, and other soft-bodied prey. Adults are attracted by the odor of aphid honeydew and lay their eggs near aphid colonies.

The Buzz on Pesticides

A *pesticide* is a substance that kills, stupefies, repels, or inhibits feeding on or prevents the infestation or destruction of a plant, place, or thing by insects, animals, or other plants. Pesticides work by modifying the physiology of a plant or the pest or by altering its natural development, productivity, quality, or reproductive capacity. Pesticides are also used to modify an effect of other chemical products and can be made to attract a pest for the purpose of destroying that pest. Pesticides control pests or other organisms by physically, chemically, or biologically interfering with their metabolism or normal behavior. Most pesticides are lethal to target pests when applied at the rate specified on the pesticide label. Some pesticides, however, are nonlethal to pests. These include repellents or attractants, sterilizing agents, defoliants, and some products that complement the action of another pesticide without being particularly toxic themselves.

Some Groups of Chemical Products Considered to Be Pesticides

Bactericides

These destroy, suppress, or prevent the spread of bacteria. Examples include swimming pool chemicals containing chlorine and products used to control scum. Household disinfectants and some industrial disinfectants are also bactericides but are not considered pesticides.

Fungicides

These control, destroy, or make harmless the effect of fungi, molds, and mildews.

Genetically Modified Organisms (GMOs)

GMOs are sometimes controversial agricultural crops that are genetically modified to make them more resistant to pests and diseases or tolerant to certain herbicides. Genetically modified products are subject to an assessment and registration process that again is greatly debated.

Herbicides

These destroy, suppress, or prevent the spread of a weed or other unwanted vegetation.

Insecticides

These destroy, suppress, retard, or inhibit the feeding of or prevent infestations or attacks by an insect. Insecticides are used to control a wide variety of insect pests, including roaches, aphids, moths, fruit flies, and locusts to name just a few. Insecticides include products such as flea powders and liquids used externally on animals.

Lures

These are chemicals that attract a pest to a pesticide for the purpose of its destruction. Pheromone lures draw in pests with the false promise of species reproduction. Food-based lures such as cheese in a mousetrap are excluded and are not considered pesticides, just a last meal.

Repellents

These repel rather than destroy a pest. Included in this category are personal insect repellents used to repel biting or disease-spreading mosquitoes and other insects.

The method we choose to deliver a pesticide to a pest depends on the nature and type of the pesticide we want to use, how it is going to be applied, and in what type of environment it will be used. Application methods include spraying, fumigating, and baiting. Most households use pesticides that are called *contact pesticides*. To be effective, contact with the pest is necessary; it must be absorbed either through the external body parts of the target insect or the exposed surfaces of a plant. Other pesticides are called *systemic pesticides*. These agents can be moved or carried from the site of application to another site within the plant, insect, or animal where they become effective. These insecticides are absorbed by foliage and spread throughout the plant, where they kill chewing or sucking insects or are transferred to the roots of the plant to kill worms or caterpillars that are feeding there. Similarly, baits take effect once they have been transferred from the digestive system to the pest's bloodstream or nervous system. Baits are meant to be ingested as a potential food source or during grooming; they are used most often for common pests in our homes such as rodents and cockroaches.

Pesticide Usage Tips

- Avoid using pesticides when alternatives are available, especially if you are pregnant.
- If you decide to use pesticides, read the labels to select the appropriate pesticide for your problem.
- Do not buy more than you can use in one or two applications.
- Do not mix pesticides unless directed to do so by label directions, and use the exact amounts specified.
- Do not wear soft contact lenses when applying pesticides; they can absorb the pesticides.
- Keep children and pets away from all areas where pesticides have been applied.
- When applying more than a squirt of pesticide, wear clothing that covers all exposed skin, chemical splash goggles, a respirator with the correct cartridge and filter, and heavy rubber or nitrile gloves.
- After using a pesticide, wash your hands and exposed skin areas before eating.
- Wash pesticide-contaminated clothing separately from other clothing.
- When a room is treated with a pesticide, leave the room for as long as recommended by the applicator or label. On returning, open all windows, and allow the room to air out. Wash contaminated surfaces.

Pesticide Storage Tips

- Always store unused pesticides in their original containers. Store inside a sealed plastic container or a metal container with a lid.
- Clearly label all containers.
- Do not store pesticides near food.
- Store in a secure area away from children and pets.
- Do not store metal containers in wet areas, on concrete floors, or in other locations that will encourage the metal to rust.

Pesticide-Free Pest-Control Tips

Remove Water

All living things need water for survival. Fix leaky plumbing, and do not let water accumulate anywhere in or around your home. For example, do not leave any water in trays under houseplants, under the refrigerator, or in the sink overnight. Remove or dry out water-damaged and wet materials. Even dampness or high humidity can attract pests.

Remove Food

Store food in sealed glass or plastic containers, and keep your kitchen clean and free from cooking grease/oil. Do not leave pet food in bowls on the counter or floor for long periods of time. Put food scraps or refuse in tightly covered, animal-proof garbage cans, and empty your garbage frequently.

Find and Fix Hiding Places

Caulk cracks and crevices to control pest access. Bathe your pets regularly, and wash any surfaces pets lie on to control fleas. Avoid storing recyclables such as newspapers, paper bags, and boxes for long periods of time. Always check for pests in packages or boxes before carrying them into your home.

Block Pest Entryways

Install screens on all floor drains, windows, and doors to discourage crawling and flying pests from entering your home. Make sure that any passageways through the floor are blocked. Place weatherstripping on doors and windows. Caulk and seal openings in walls. Keep doors shut when not in use.

Remove or Destroy Hiding Places

Remove piles of wood from under or around your home to avoid attracting termites and carpenter ants. Destroy diseased plants, tree prunings, and fallen fruit that may harbor pests. Rake fallen leaves. Keep vegetation, shrubs, and wood mulch from touching your house.

Remove Breeding Sites

Clean up pet droppings from your yard that can attract flies that spread bacteria. Pick up and cover litter or garbage that can draw mice, rats, and other critters. Drain off, fill in, or sweep away standing-water puddles that can be a breeding place for mosquitoes. Make sure that drain pipes and other water sources drain away from your house, and keep them free of debris.

Take Care of Plants

This includes your lawn, flowers, fruit and shade trees, vegetables, and other plants. Good, strong plants resist pests better than weak plants. Use mulch to reduce weeds and maintain even soil temperature and moisture. Water adequately. Native flowers, shrubs, and trees are often the best choices because they require minimal care.

Choose a Type of Grass That Grows Well in Your Area

If you have very little rain or poor soil, don't plant a type of grass that needs a lot of water or extra fertilizer.

Cut It High and Let It Lie

Grass that is a little longer than normal makes for a strong, healthy lawn with fewer problems. Weeds have a hard time taking root and growing when grass is fairly tall. A foot-high meadow isn't necessary for most of us; just try adding an inch to the length of your grass.

Pest-Control Goals

It is impossible to get rid of all weeds and pests. A healthy lawn probably will always have some weeds and some insect pests. A healthy lawn also will have friendly insects and organisms such as earthworms, and pesticides also can kill these beneficial friends.

CHAPTER 5

Recycling Plastics and Composite Projects

THE KINDS OF PLASTICS found in our homes and how can we recycle them depend on what the plastics are made from and the types of characteristics they have. What are plastics? Plastics and composites are made of organic materials such as carbon (C), hydrogen (H), nitrogen (N), chlorine (Cl), and sulfur (S) that have properties similar to those of materials found in nature, such as wood. Organic materials are based on polymers, which are produced by the conversion of natural substances or by chemical reactions involving substances such as oil, natural gas, or coal. The chemical reaction of these monomers causes them to bond into chains called *polymers*. Different combinations of monomers yield plastic resins with different properties and characteristics. The resulting resins may be molded or formed to produce several different kinds of plastic products for use throughout our homes. A resin can be tailored to a specific design or performance requirement. This is why certain plastics are best for certain specific applications, whereas others are best for entirely different applications. For example, plastics can be designed for increased impact strength and the ability to withstand shock loading; some plastics have increased heat resistance to protect them from exposure to higher temperatures or increased chemical resistance to protect against breakdown in harsh chemical environments. Even though the basic makeup of many polymers is simply carbon and hydrogen, other elements are added to impart various qualities. Oxygen, chlorine, fluorine, nitrogen, silicon, phosphorous, mineral fillers, and sulfur all can be found in the molecular makeup of various polymers. Polyvinyl chloride (PVC) contains chlorine. Nylon contains nitrogen. Teflon contains fluorine. Polyester and polycarbonate plastics contain oxygen.

Two Characterizations of Plastic

A *thermoset* plastic is a polymer that solidifies, or "sets," irreversibly when heated. Thermoset polymers can't be softened again once they set. Thermosets are valued for their durability and strength and are used primarily in automobiles and construction, although they can undergo such applications as adhesives, inks, and coatings.

Other examples of thermoset plastics and their product applications include

- *Polyurethanes*—mattresses, cushions, insulation, car bumpers, and toys
- *Unsaturated polyesters*—lacquers, varnishes, car bodies, boat hulls, and furniture
- *Epoxies*—glues, coatings for electrical circuits, helicopter blades, and aircraft skins

A *thermoplastic* is a polymer in which the molecules are held together by weak secondary bonding forces that soften when exposed to heat and return to the original condition when cooled back down to room temperature. When a thermoplastic is softened by heat, it then can be shaped by extrusion, molding, or pressing. Thermoplastics offer versatility and a wide range of applications and are used most often around our homes because they can be rapidly and economically formed into any shape needed to fulfill the packaging function. Examples include milk jugs and soda bottles.

What Is What?

To understand what kinds of plastics we have around us every day and what types of physical characteristics they have, we need to know what the plastic's formulation is. It would be nice if plastics of different types would mix together easily for recycling, but they often don't because melt temperatures and even material compatibilities vary so much. To address this issue, a code was developed to help consumers and industry easily separate plastics for ease of recycling. The group that instituted the plastic coding system is called the *Society of the Plastic Industry* (SPI). The system has been used as a resin identification code for over 20 years, and it was created at the urging of recyclers around the country. A growing number of communities were implementing recycling programs in an effort to decrease the volume of waste subject to tipping fees at landfills. The SPI code was developed to meet recycling processors' needs while providing manufacturers with a consistent, uniform system that could be applied nationwide. Some recycling processors may require that the plastics be sorted by type and separated from other recyclables, whereas others specify that mixed plastics are acceptable if they are separated from other recyclables.

Ode to the Code

The resin ID code can be found on the bottoms of most plastic containers and packaging items. Most of the plastics in our homes that are recyclable are from packaging. Code numbers range from 1 to 7. The following paragraphs provide a breakdown of the resin ID numbers, common types of plastics and their most common household uses, what the different plastics are recycled into, and a few of their physical characteristics.

TIP **Look at the recycling number on the bottom of the plastic container. The safest plastics for food-related uses are numbers 1, 2, 4, and 5.**

Number 1 Plastic: Polyethylene Terephthalate (PET or PETE)

PET plastic is used most commonly for single-use beverage bottles because it is inexpensive, light in weight, and easy to recycle. It is a crystal clear virgin plastic that's semirigid (depending on thickness), and it has a fairly high melting point, as well as good impact resistance under normal to cool conditions. It poses a low risk of leaching breakdown products.

PET is used to make soft-drink, water, and beer bottles; mouthwash bottles; peanut butter containers; salad dressing and vegetable oil containers; and even "ovenable" TV dinner trays. It is recycled into polar fleece, fiber, tote bags, furniture, carpet, paneling, straps (occasionally), and new PET containers. However, recycling rates remain relatively low (around 20 percent) even though the material is in high demand by manufacturers.

Number 2 Plastic: High-Density Polyethylene (HDPE)

HDPE is a versatile plastic with many uses, especially for packaging. It is a semiclear virgin plastic that's semiflexible (depending on thickness), and it has a midlevel melting point and fair impact resistance in cool temperatures. It carries a low risk of leaching and is readily recyclable into many goods. HDPE is used to make milk jugs; juice bottles; bleach, detergent, and household cleaner bottles; shampoo bottles; some trash and shopping bags; motor oil bottles; butter and yogurt tubs; and even cereal box liners. It is recycled into laundry detergent bottles, oil bottles, pens, recycling containers, floor tile, drainage pipe, lumber, benches, picnic tables, fencing, and even doghouses.

Number 3 Plastics: Vinyl (V) and Polyvinyl Chloride (PVC)

PVC is a tough plastic that weathers well, so it is used commonly for piping, siding, and similar applications. In addition, PVC is used to make window cleaner and detergent bottles, shampoo bottles, cooking oil bottles, clear food packaging, electrical wire covering, medical equipment, and window frames. It is a nontransparent to clear virgin plastic depending on its additives and intended use. It is also flexible or rigid, again based on its additives and desired use. It has a low melting point unless mixed with a mineral such as calcium, making CPVP. However, these plastics have very little impact resistance in cold temperatures. PVC contains chlorine, so its manufacture can be highly dangerous. It is recycled into decking, fencing, paneling, mudflaps, roadway gutters, flooring, sewer pipe, speed bumps, and even floor mats.

CAUTION **Never burn PVC, and if it is burning, leave the area immediately because it releases toxins as it burns.**

Public Enemy Number 1

PVC or vinyl poses a particular hazard to health and the environment. The problems with PVC include dioxin emissions when it is manufactured or burned and the effects of additives such as phthalates, which are called *plasticizers*. Phthalates allegedly mimic the body's hormones and may cause reproductive and neurologic damage.

CAUTION **When PVC is overheated or burned, it gives off toxic hydrogen chloride gas, which turns into hydrochloric acid on contact with moisture within the lungs. Don't try to recycle this plastic in your home; it's just too dangerous. Find a facility that accepts PVC for recycling instead.**

Number 4 Plastic: Low-Density Polyethylene (LDPE)

LDPE is a flexible plastic with many applications, such as squeezable bottles, bread wrappers, frozen-food trays, dry-cleaning bags, shopping bags, tote bags, clothing, furniture, reusable storage containers, and carpets. It is a semiclear virgin plastic that is semiflexible (depending on thickness), and it has a midlevel melting point and better impact resistance in cool temperatures than high-density polyethylene (HDPE). It is recycled into trash can liners, trash cans, compost bins, shipping envelopes, paneling, lumber, landscaping ties, floor tile, and roofing materials.

Number 5 Plastic: Polypropylene (PP)

PP has a high melting point, so it is often chosen for containers that must accept hot liquids. It is used to make yogurt containers, syrup bottles, ketchup bottles, bottle caps, straws, medicine bottles, and automotive battery housings. It is a semiclear virgin plastic that is semiflexible (depending on thickness), and it has a high melting point and better impact resistance in cool temperatures than PET. It is recycled into rope, brooms, brushes, automobile battery cases, ice scrapers, landscape borders, bicycle racks, rakes, bins, and freight pallets for shipping.

Number 6 Plastic: Polystyrene (PS)

Polystyrene can be made into rigid or foam products. It is popularly known best as the trademark *Styrofoam*. It is a nontransparent virgin plastic depending on its additives and intended use, and it is also semiflexible or rigid, again based on its additives and desired use. When gases such as carbon dioxide (CO_2) are added to it, heated, and allowed to expand, the result is a plastic foam. It has a low melting point (unless mixed with calcium) and has very little impact resistance in cold temperatures in its rigid state. However, it does well with impact in the foamed state regardless of the temperature, and it is a very effective insulation. This material was long on environmentalists' hit lists for dispersing widely across the landscape and for being notoriously difficult to recycle. It is used in disposable plates and cups, meat trays, egg cartons, carry-out containers, computer housing, and even compact disc cases. It is recycled into insulation, light switch plates, egg cartons, vents, rulers, foam packing, carry-out containers, and toys.

CAUTION **Polystyrene containers can leach out chemicals into food (when heated in a microwave). It is suggested that these chemicals can threaten human health and reproductive systems. Don't heat foods in foam-type plastics if you're at all worried about it.**

Number 7 Plastics: Miscellaneous

A wide variety of plastic resins that don't fit into the previous categories are lumped into number 7. A few are even made from plants and are compostable. These plastics are used in polycarbonate (Plexiglas) sheeting, large water bottles, bulletproof materials, sunglasses, DVDs, computer parts, signs and displays, certain food containers, and nylon. Number 7 also covers plastics that are layered together for a specific use, such as a squeezable catsup bottle has an inner PET liner that doesn't leach out into the ketchup and a flexible LDPE outer cover that allows it to perform as the consumer expects. These plastics are recycled into plastic lumber and custom-made products. Number 7 can be any type of plastic in any type of arrangement, and it is hard to recycle.

Public Enemy Number 2

Polycarbonate plastic also has been identified as being potentially risky because it may have undesirable effects on the foods packaged in it. Polycarbonate plastics may contain a substance called *bisphenol* (BPA) that mimics the effects of estrogen in the body, causing undesired health and development issues, especially in the young.

Risk Management

Consider the type of plastic when you buy the following items, most made from PVC: baby toys, bibs and chew toys, cling wraps and food containers, and children's lunch boxes. Baby bottles and water bottles are often made from polycarbonate. Look for cling wraps, baby items, and other products with labels such as "No PVC," "No chlorine," or "No plasticizers."

TIP **Microwave food in glass or ceramic containers with no metallic paint. Don't heat plastic containers or plastic wrap in the microwave.**

Board with Plastics

Most plastic lumber products on the market today are made from a single resin, polyethylene, which is available in high and low densities (HDPE and LDPE). All plastic types or plastic resins currently used for making plastic lumber share a common origin in fossil fuels. Fossil fuels are the decayed remains of millions of years of organic and microbial life. Gravity, heat, and pressure from the earth are how oil, natural gas, and coal came to be. Plastics come from a common source that is known to have a negative impact on the environment as well as public health. Aside from the fact that all plastics use fossil fuels as a common building material, they differ greatly in other respects. Depending on the manufacturing process and the additives used in formulating plastic products, some pose greater chemical hazards in their production, use, and disposal than others. While no plastic is environmentally benign, polyethylene poses less risk in terms of environmental or health impacts, making it preferable to plastics that have pose greater risks such as polystyrene and polyvinyl chloride. To improve structural performance qualities such as rigidity and strength, some plastic lumber manufacturers reinforce the primary plastic resin with other materials. Fiberglass is one material option used often to increase the load-bearing capacity of plastic lumber. The use of wood fiber, minerals, and other solid additives helps to lower material costs and alter the properties of the plastic lumber the end user may want.

Products Made with Recycled Plastics

- Dimensional lumber
- Decking
- Boardwalks and walkways
- Marine docks
- Fencing and posts
- Picnic tables
- Benches
- Bridges
- Retaining walls
- Railroad ties
- Pallets
- Planters and landscaping timbers
- Trash can receptacles
- Playground equipment
- Compost bins
- Animal stalls
- Sound barriers
- Parking stops
- Signposts and signs
- Bicycle racks
- Bricks

Project 1
Make Your Own Wafer Board-Style Plastic Lumber

This is a product my family and I used to make over 15 years ago. I hope you enjoy it. This is much like the wooden-style wafer board called *oriented strand board* (OSB) you can see at lumber yards and in use on buildings under construction as flooring and sheathing. We can make a material that is like the wooden counterpart but that is stronger, lighter, and more resistant to the elements that quickly destroy wood. As our plastic wafers soften and begin to melt together, the lighter and lower-melting plastics are acting as a thermosetting glue that bonds all the pieces together and sets up hard once cooled. We add baking soda to it to release CO_2 gas as it is heated, and this causes our board to be porous and strong (like a honeycomb structure), and the CO_2 also discourages oxygen's weakening effects on our plastic's chemical bonds. As the plastic softens from being heated, the pieces start to shrink and curl together as bonds are made. We just want to melt the edges of the plastic wafers to make them stick and bond together; we don't want to bring the plastic to a full melt.

Mass-Production Method

In an industrial setting, we would make this plastic lumber by using rotary shears to size our plastic pieces, ribbon blenders to mix in the sodium bicarbonate, an infrared conveyor oven to heat the plastic, and a roller press to compress it.

WHAT YOU'LL NEED

- A trash can full of mixed plastics with the code numbers 1, 2, 4, 5, and 6
- A roll or large piece of recycled aluminum foil
- One box of baking soda (sodium bicarbonate)
- A few drops of vegetable oil
- A piece of wooden lumber or a stick that's just large enough for you to hold comfortably
- A pair of construction-style scissors, razor knife, or hacksaw
- An old cookie sheet or cake pan
- A pair of heat-resistant gloves
- An electric kitchen oven
- Safety goggles (Figure 5-1)

TIP You will want at least half the plastic for this project to be bag or film plastics, both of which are most often coded 2 and 4.

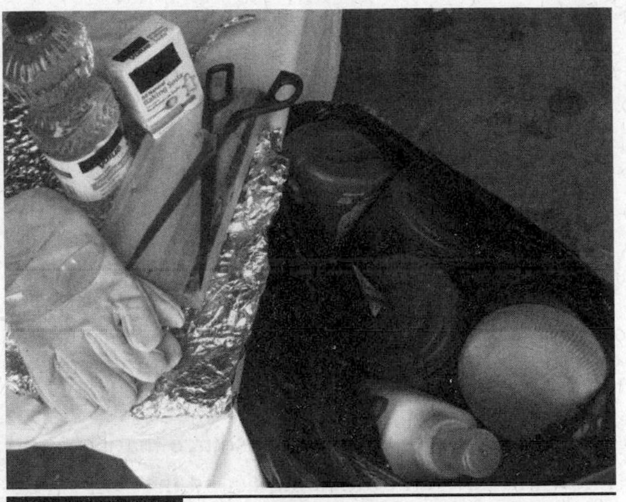

Figure 5-1

Let's Start

TIP Any time you work with plastics and heat is involved, it is very important not to overheat the plastic or allow it to catch on fire. When plastics are burned, they can produce hazardous gases and fumes such as chlorine gas, dioxins, and carbon dioxide. Plastics also are very hard to extinguish once ignited and can burn and get out of control easily. Proper fire extinguishers should be available in every area where people work. If you don't have fire protection, please do not attempt any of the projects in this book that require heat. And *always* wear eye protection

1. First, we need to make the plastic wafers.
2. Cut the plastic into pieces the size of an average cookie or smaller. Take your time, wear your gloves, and use whatever allows you to comfortably cut the plastic into wafers (Figure 5-2).

TIP Plastic bags and film cut very easily with scissors and can be used exclusively for this project if you find the thicker plastics difficult to size. The tops of bottles that are thick can be set aside for other projects later. If you have access to a band saw, it can make quick work of cutting the more rigid plastics.

Figure 5-2

3. Preheat your oven to 375°F (190°C).
4. Line your cookie sheet or cake pan with foil, and rub a little vegetable oil on the foil.
5. Now sprinkle a dusting of baking soda on the foil (Figure 5-3).
6. Layer about a half inch of plastic wafers on top, and repeat the dusting of baking soda (Figure 5-4).
7. Repeat these steps until you have about a finger's depth of material (Figure 5-5).
8. End with a dusting of baking soda as the final layer.

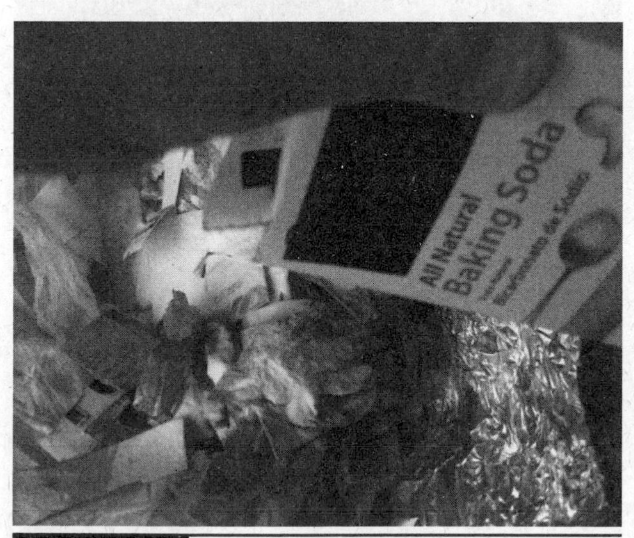

Figure 5-3

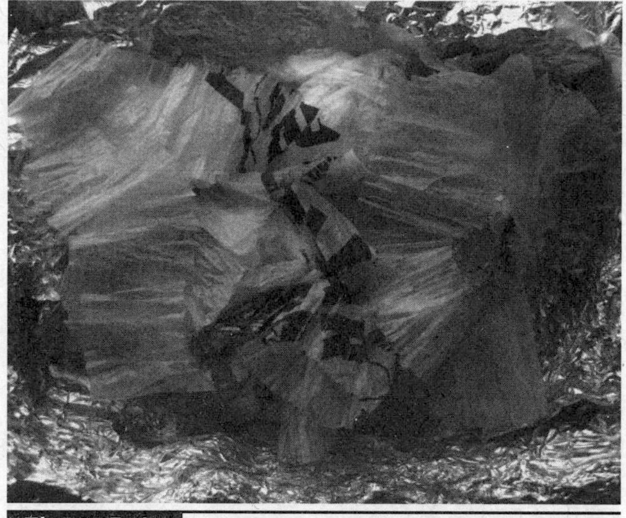

Figure 5-4

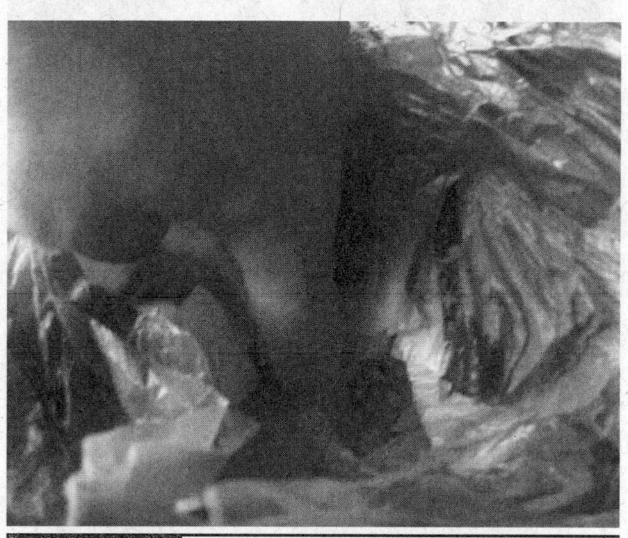

Figure 5-5

9. Once the oven is preheated, place the wafers on the middle rack in the center of the oven, and let the material heat for about 30 minutes.
10. Open the oven, and carefully look to see if the plastic wafers have shrunk or curled some. If not, increase the temperature a little, and "cook" for another 15 minutes.
11. Wearing your heat-resistant gloves, carefully remove the melted wafers, and place them on a heat-safe surface. With your heat-resistant gloves still on, take the piece of handheld wood and press the melted wafers down to

about ¾ inch (19 mm) thick. If the material doesn't stay compressed, you need to put it back in the oven for another 15 minutes and repeat if needed (Figures 5-6 and 5-7).

12. Once compressed, allow the material to cool to room temperature, and then remove it from the foil. This is your first piece of wafer board–style plastic lumber.

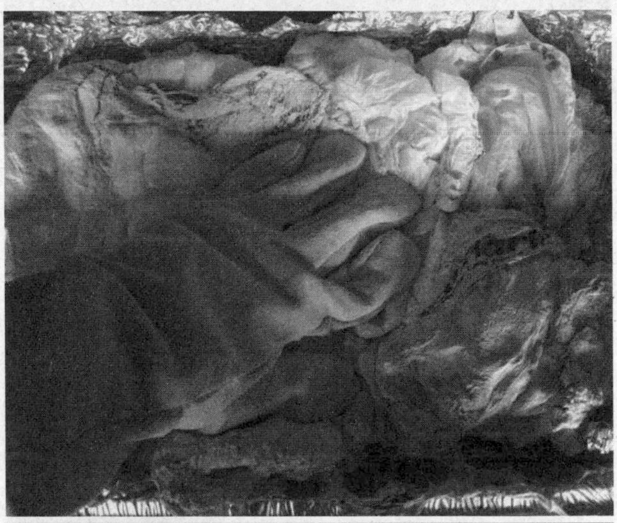

Figure 5-6

Figure 5-7

Project 2
Make a Square Panel Out of Your Plastic Wafer Board

If you plan on building something with your plastic wafer board, it first must have square, straight edges so that you have proper angles and can get accurate measurements.

WHAT YOU'LL NEED

- Rough-edged plastic wafer board
- A yardstick or straight edge
- An adjustable square
- A tape measure
- Safety goggles
- A table saw
- A pencil (Figure 5-8)

Figure 5-8

Let's Start

TIP Always use appropriate safety equipment, such as goggles, earplugs, and dust masks. Do not wear gloves when working with most tools. Always wear eye protection when working with metal. Do not wear sandals, open-toed shoes, or canvas shoes when working with tools. Avoid loose-fitting clothes that might become entangled in a power tool. Remove rings and other jewelry. Read the owner's manual before using any tool. Never use a tool unless you are trained to do so. Inspect each tool before each use, and replace or repair if parts are worn or damaged.

1. First, we need to make a straight edge on our plastic wafer board so that we can square it up. With your yardstick or straight edge, draw a line that is flush with the longest edge of your piece (Figure 5-9).
2. Wear eye protection, and with your saw, carefully cut along the line you just drew. The first cut is the hardest, so go slow and pay great attention. Use your miter gauge to help guide you (Figure 5-10).
3. Once the board is cut, use your square to draw a 90-degree line that is flush to the edges of the board (Figure 5-11).

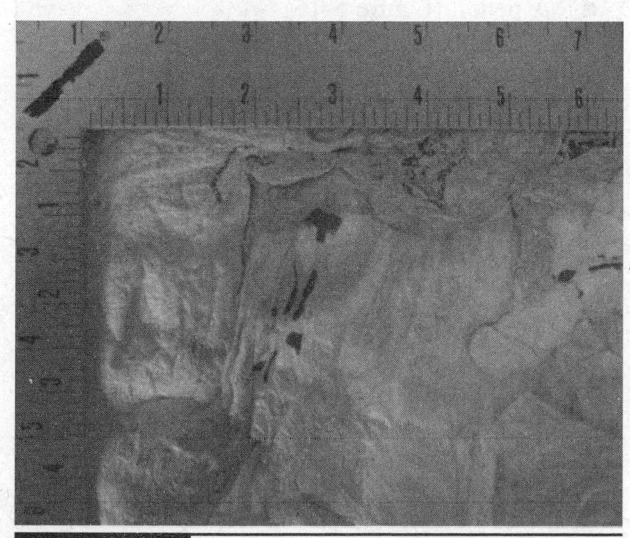

Figure 5-9

Figure 5-10

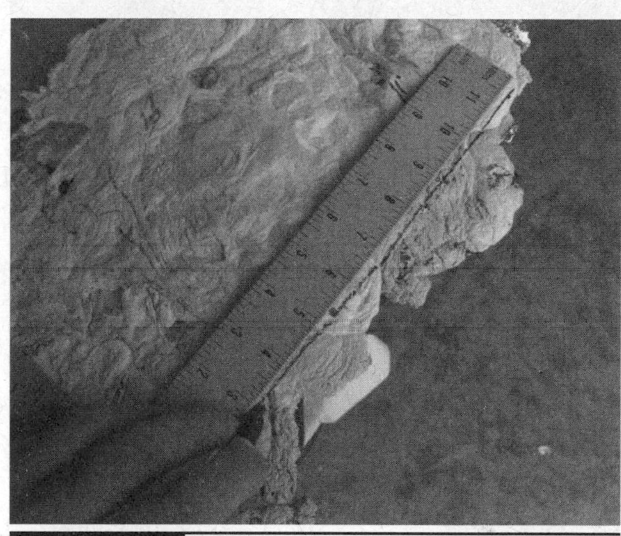

Figure 5-11

4. Now cut those lines with your saw. You can use the miter gauge to help you do this. If you have a large enough table saw, you use your table saw's fence and miter for the last two cuts if you have a square 90-degree corner with which to work (Figures 5-12 and 5-13).
5. Now use your square again to make a 90-degree line from those last two cuts flush along the last rough edge of your plastic wafer board. If you need more of these pieces, repeat steps 1 through 5 (Figure 5-14).

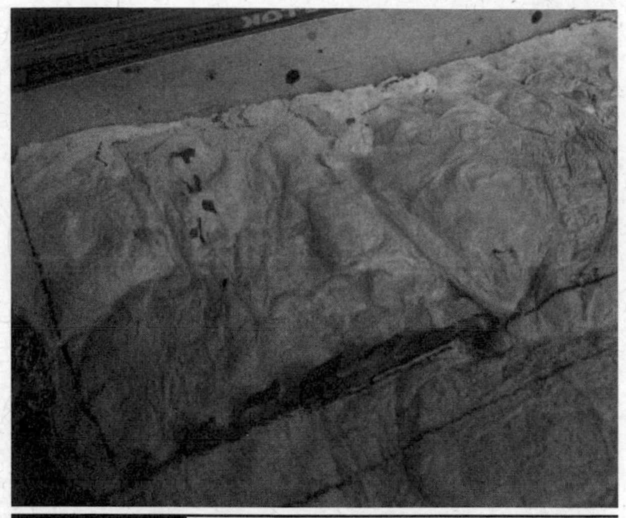
Figure 5-12

Figure 5-13

Figure 5-14

Project 3
Make a Birdhouse Out of Plastic Lumber

A birdhouse is a good way to bring nature closer to your home, and it's nice to be able to enjoy nature from the comfort of your home, as well as contributing to the environment by providing an essential place for birds to nest (Figure 5-15).

WHAT YOU'LL NEED

- Five pieces of ¾- × 9- × 18-inch plastic lumber
- Eleven 4-inch decking screws
- A small drill bit to make pilot holes for decking screws
- A Phillips drive for the decking screws
- A drill or cordless screwdriver
- A yardstick or straight edge
- An adjustable square
- A 1½-inch drill bit
- A tape measure
- A ⅜-inch drill bit
- Safety goggles
- A table saw
- A pencil (Figure 5-16)

Figure 5-15

Figure 5-16

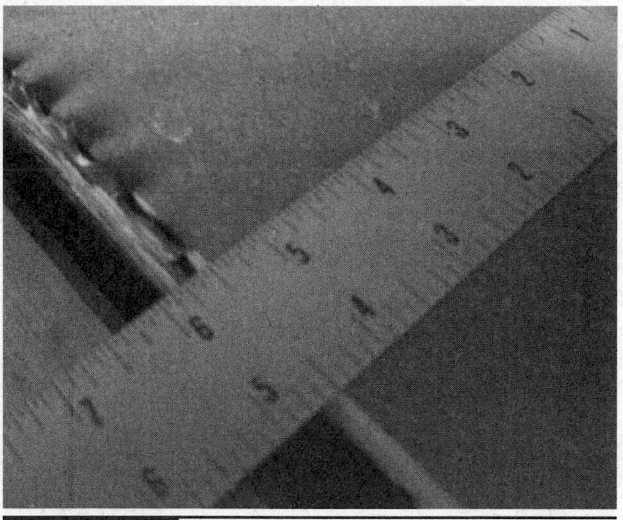
Figure 5-17

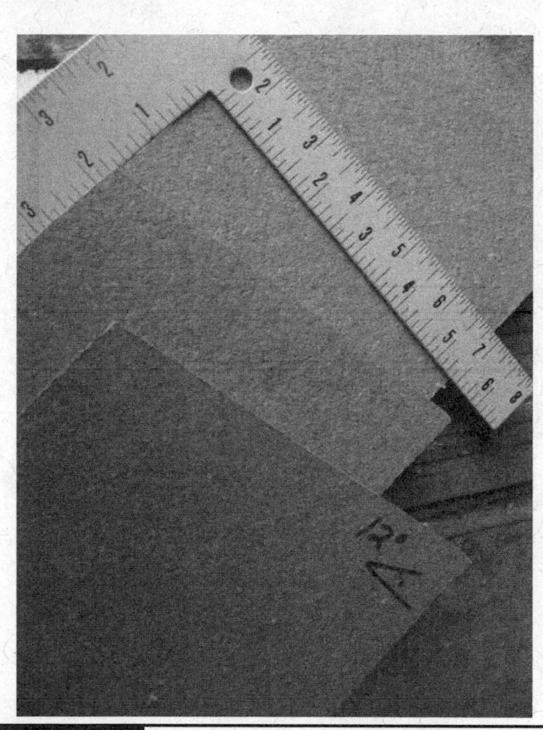
Figure 5-18

Let's Start

1. Let's make the back of the birdhouse, Cut a piece with your table saw that measures 18 × 5½ inches with a 12-degree bevel cut on the top edge (Figure 5-17).
2. Now let's make the sides. Cut two pieces that measure 11½ × 5½ inches with a 12-degree angle cut (Figure 5-18).
3. To make the floor, measure and cut a 4- × 5½-inch piece, and then cut about ⅜ inch from the corners to allow for airflow (Figure 5-19).
4. Let's make the front, so measure and cut a 10½- × 5½-inch piece with a 12-degree bevel cut on the top edge. Cut a centered 1½-inch entrance hole 2 inches from the top (Figure 5-20).
5. To make the roof, measure and cut a 9- × 8-inch piece. The 9-inch length will face the front; the 8-inch edge goes with the width (Figure 5-21).

Figure 5-19

Figure 5-20

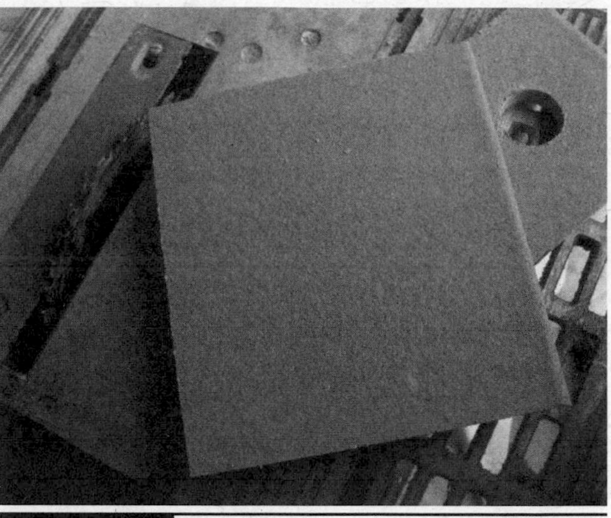

Figure 5-21

Figure 5-22

6. Screw the sides to the back. You may have to drill pilot holes if you can't make the screws go in straight (Figure 5-22).
7. Now recess the floor about ½ inch from the bottom opening, and screw it into place from the sides (Figure 5-23).
8. On the inside of the front just below the entrance hole, cut about 5 grooves that are ⅛ inch deep and ½ inch apart. Use your table saw by lowering the blade to ⅛ inch and using your miter gauge (Figure 5-24).
9. Align the front with the tops of the sides. Measure 1 inch from the top of the side, and drill a pilot hole on the left and right sides. Then add decking screws through the holes to keep the screws from being too snug. You want to allow the lid to swing open for cleaning (Figure 5-25).
10. Center the roof, and screw it down. The front and sides will have overhangs, whereas the back sits flush (Figure 5-26).

Figure 5-23

Figure 5-24

Figure 5-25

Figure 5-26

Figure 5-27

11. Use decking screws through the back of the birdhouse to attach it to a tree, post, or wall (Figure 5-27).

How to Use It

After you make your birdhouse, you also should consider the types of birds that you want to attract and make sure that the birdhouse is suitable for the bird you want. You should know various types of birds in your local area. Attach the birdhouse in the best location to a tree or post with your decking screws by holding open the front and going through the back. Remember to clean out the birdhouse at year's end for next year's birds. You can prepare special seeds for foreign birds if you know for sure that your area is part of their migration route. It will be interesting to watch these new birds for a change after you have watched the common ones year after year. Go to a local pet store to find the right bird seeds to attract the birds you want. If you have placed the birdhouse in an appropriate location and fill the bird feeder with enough seeds, then it is time to relax and enjoy the sight of the birds that visit.

Project 4
Make a Recycled-Plastic Landscape Block

Mixing different kinds of plastics to recycle them into a usable item can be very tricky or just plain impossible. Years ago, while I was testing blends of plastics, I learned that mixing some plastics worked well if there was some kind of mineral or organic matrix for the plastics to bond to. You'll find a lot of the plastic lumber on the market today uses wood waste in the form of sawdust or wood flour. This wood matrix is used as a filler to help lower material cost as well as give the plastic something to bond to more aggressively than just the plastic itself. In this project, you can use pea gravel, sand, or finely pulverized glass that is free of dust. You heat the pea gravel with a small amount of vegetable oil, which helps it convey heat energy and acts as a solvent to help melt the plastics. You are basically heating the mineral (rock) and oil mixture to a temperature that is hot enough to reach to melt the plastics you are adding to it. The result is like a plastic marshmallow crispy treat of sorts.

The Plastic Recipe

The plastic types we can use for this project are all types except number 3, PVC, which is dangerous to heat and also has too low of a melting point for our results. Number 1 (PETE/PET) and number 5 (PP) plastics have too high a melting point, but they can be used as a filler of sorts if small enough pieces are added to the blend. You want your commingled block to be a little over three-quarters (75 percent) number 2 HDPE and number 4 LDPE. Filler plastics include number 1 PETE/PET, number 5 PP, number 6 PS, and any number 7 plastic and can make up the remaining 25 percent of the mix if the pieces are small. Plastics such as number 6 PS and number 7 still may become part of the melt.

Mass Production

In an industrial setting, we would make these comingled plastic blocks by using a ribbon blender to mix the heated mineral with the plastics. The melt mix then would be partially extruded into a mold carousel to quickly make the profiles of the products we want.

TIP **You will need at least three-quarters of the plastic for this project to be from number 2 HDPE and number 4 LDPE.**

WHAT YOU'LL NEED

- A couple kitchen trash cans full of mixed plastics with code numbers 2, 4, 5, 6, and 7
- A few drops of vegetable oil
- A 1- × 4-inch × 5-foot board
- Clean, dry sand and pea gravel
- A good-sized stick or old utensil with which to mix, stir, and scrape out the pot
- A electric hot plate or gas burner that you can use outside
- A old 8-quart or larger pot with a handle(s)
- A drill or cordless screwdriver
- A pair of heat-resistant gloves
- A tape measure
- Safety goggles
- A razor knife
- A pencil or pen (Figure 5-28)

Figure 5-28

Let's Start

TIP Any time you work with plastics and heat is involved, it is very important not to overheat them or allow them to catch on fire. When plastics are burned, they can produce hazardous gases and fumes such as chlorine gas, dioxins, and carbon dioxide. Plastics also are very hard to extinguish once ignited and can burn and get out of control quickly. Proper fire extinguishers should be available in every area where people work, and they should be able to put out various types of fires. If you don't have fire protection, please do not attempt any of the projects in this book that require heat. And always use the appropriate safety equipment, such as goggles, earplugs, and dust masks. Do not wear gloves when working with most tools. Always wear eye protection when working with metal. Do not wear sandals, open-toed shoes, or canvas shoes when working with tools. Avoid loose-fitting clothes that might become entangled in a power tool. Remove rings and other jewelry. Read the owner's manual before using any tool. Never use a tool unless trained to do so. Inspect the tool before each use, and replace or repair it if parts are worn or damaged.

1. Take a 1- × 4-inch × 5-foot board and crosscut it into two 6-inch (15-centimeter) pieces and three 16-inch (40-centimeter) pieces (Figure 5-29).

Figure 5-29

2. Take the three 16-inch (40-centimeter) pieces and screw them together with the bottom edges flush with the outside edges of another 16-inch (40-centimeter) board, making a U shape (Figure 5-30).
3. Now take the two 6-inch (15-centimeter) pieces and screw them to the ends of the U shape, making a box and your landscape block mold (Figure 5-31).
4. Take the mold and coat it with a thin layer of vegetable oil to use as a mold release.
5. Place your pot on an outdoor burner in the open air.
6. Measure the amount of gravel you need for your mold, and empty it into the pot. Now add just enough vegetable oil to make the pea gravel look like it's wet (Figures 5-32 and 5-33).
7. With gloves on, turn on the heat, and add just a few small pieces of the lower-melting plastics to the gravel and oil mixture.
8. As the mixture warms, use the stick to move the pieces of plastic around.
9. As soon as the plastic softens and begins to melt (turn the heat down if you see or smell smoke), start adding more of the plastic while

Figure 5-30

Figure 5-31

Figure 5-32

Figure 5-33

stirring, and keep adding plastic until you have enough to completely cover all the gravel with molten plastic (looks like a chocolate rice marshmallow treat) (Figures 5-34 and 5-35).

10. Turn off the burner when you are ready to make the landscape block.
11. With your gloves on and using your stick to help, carefully pour the hot mix into the mold, filling any cavities you may see (Figures 5-36 and 5-37).
12. Let the block cool, and then turn the mold upside down, tapping the mold opening on a solid surface to release the block from the mold (Figures 5-38 and 5-39).
13. Repeat the preceding steps to make as many blocks as you want.

Figure 5-34

Figure 5-36

Figure 5-35

Figure 5-37

Figure 5-38

Figure 5-39

Project 5
Make Recycled-Asphalt-Shingle Paver Bricks

As the name implies, this is a block made of waste roofing shingles. Years ago, my family patented and manufactured blocks such as these. Asphalt roofing waste is plentiful, and I thought that this project would be a good one to share.

Background

This project involves converting waste asphalt shingles into useful dimensional construction materials, including and without limitation to paving blocks for walkways or patios or landscape borders for flower or shrub gardens, retaining wall blocks, bricks, tire stops, curbs, tiles, posts, beams, and so on. Asphalt roofing shingles, needless to say, are in widespread use across the country on structures of all types. What to date has not been a sensational or headline-worthy fact about asphalt roofing shingles is that they are responsible for a massively voluminous construction waste stream. This waste stream is, generally speaking, fed by two sources. Most of the waste stream consists of spent shingles that are removed when roofs are replaced or during building demolition. Much of the rest of the waste stream consists of scrap from new shingles used in roofing, such as the tabs that are cut out during shaping of completed shingles or discarded rejects that failed to pass quality standards.

The portion of the waste stream consisting of new shingles contains fresh asphalt. The portion of the waste stream consisting of spent shingles is likely to include a variety of contaminant debris such nails, swatches of roofing felt (an underlying and redundant vapor barrier), wood from the underlying substrate (plywood), perhaps some metallic strip material from flashing, and so on.

Spent asphalt shingles are a regulated solid waste. They must be sent to either a sanitary landfill or a special-purpose demolition landfill. For roofing contractors, there is a subeconomy involved in the disposal of spent shingles that must be dealt with and figured into the calculus of a roofing job because, if ignored, all the time, money, and equipment tied up in disposing of the spent shingles can wipe out profit. For example, after a roofing contractor has torn off a roof, he or she is now faced with disposal. The city landfill 10 miles north of the center of town accepts such debris, at a charge of $30 per ton. The contractor, depending on where the job site is, actually may have a longer trip to go than the 10 miles from the center of town. Also, the roofing contractor may be charged a surcharge for offloading if he or she cannot offload in accordance with the landfill's policies. More significantly, there is usually a wait in a line before the landfill can service the contractor's load, and at times, the wait can be significantly extended if the line is long. This waiting period is most frustrating because it idles personnel and truck(s) for the duration. This is especially true when, after a hailstorm or similar roof-damaging weather event, the roofing contractor experiences a temporary bonanza in business.

At least one containerized trash hauler has reacted to the plight of roofing contractors with a specialized service in which at least one centrally located transfer station also accepts the spent shingle debris. This often has a premium charge, naturally enough. The basic charge is more than double per ton, which excludes surcharges if help is needed offloading the debris and placing it in the container(s).

Regardless of which option the roofing contractor uses to dispose of the spent shingles, the outcome for the spent shingles is the same—they are landfilled. My realization was that this stream of waste shingles could be diverted to a processing center and recycled into useful products. Moreover, the processing center could be coordinated with one or more collection centers to service the needs of a number of roofing contractors (or whoever else finds themselves with waste shingles) and provide a conveniently located dump site, thus ensuring a continuous supply of feedstock (waste shingles) for the processing center. A number of additional advantages of this process will become apparent as you work through this project.

Summary of the Invention

The object of this invention is to recycle waste asphalt roofing shingles into useful dimensional construction products. It is an additional object of this invention to recycle waste asphalt shingles by a process that enhances the fusibility and/or flux activity of the asphalt while taking care not to overdo it at the risk of decomposing a substantial amount of the asphalt into an ashlike brittleness. Another object of this invention is to facilitate the recycling process without the need for fillers or solvents.

In the process, asphalt roofing shingles are deconsolidated into a crumble. The crumble is processed in a manner that increases the fusibility of the asphalt without decomposing a substantial amount to ashlike brittleness. This is achieved by the heat-exchange process that uses infrared heaters. In one instance, the heaters warm the bulk of the crumble to between 130 and 150°F (55 and 66°C). The power consumption to achieve this corresponds to about 65 watthours for each pound of the asphalt shingle crumble. In other words, the power consumption for each pound of crumble was as small as that consumed by a 65-watt light bulb in one hour.

Next, the asphalt shingle crumble is compacted into the shape of given dimensional construction product, which can include, without limitation, such products as retaining wall blocks, paver blocks, tire stops, bricks, curbs, tiles, posts, beams, and so on. In the specific case of paver blocks, the best quality blocks have been achieved by a two-step compaction process that includes a certain amount of elapsed time between the two compaction steps. Thus the process involves coordinating the heat source to raise the crumble to the appropriate temperature with developing the proper compaction pressure(s) and time sequence to produce the finished product.

Preferably an in-feed of asphalt composition roofing shingles is obtained by diverting waste shingles headed for landfills into processing centers. Experience shows that waste shingles include foreign debris associated with roof tear-offs, such as nails, roof felt, wood from the roof deck, pieces of flashing, and so on. Some of the foreign debris must be separated out, especially a substantial portion of the nails. The remainder is not separated out and ends up harmlessly "frozen" in the matrix of the finished product.

The deconsolidation the asphalt roofing shingles optionally can be achieved by grinding. However this is achieved, the shingles preferably are deconsolidated into a crumble that ranges between extremes of fine and coarse matter, the coarse matter preferably being chunks about ½ inch in size.

WHAT YOU'LL NEED

- A small pile of spent asphalt shingles
- A small amount of vegetable oil
- Eleven 4-inch decking screws
- A 1- × 4-inch × 4-foot-long board
- A good-sized stick or old utensil to mix, stir, and scrape out the pot
- A electric hot plate or gas burner that you can use outside
- A old 8-quart or larger pan with a handle(s)
- A Phillips drive for the decking screws
- A drill or cordless screwdriver
- A pair of heat-resistant gloves
- A table saw or miter saw
- A tape measure
- Safety goggles
- A razor knife
- A pencil/pen (Figure 5-40)

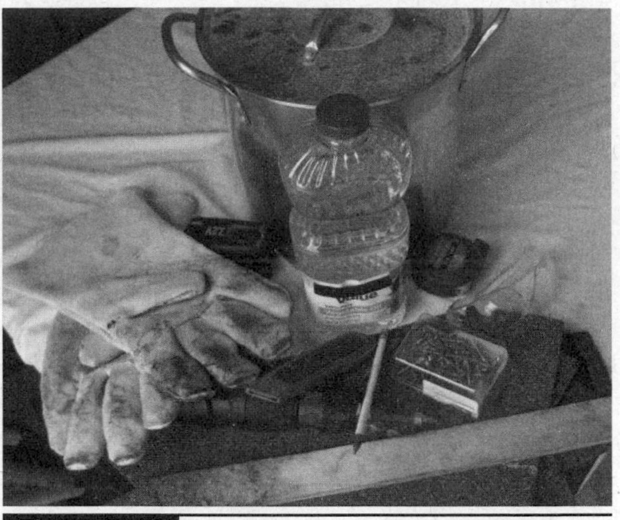

Figure 5-40

The compaction steps alternately can consist of extrusion, press-forming, or both on the same product. In the specific case of paver blocks, the press-forming is done after a certain amount of time has elapsed after a previous extrusion step.

Let's Start

TIP Any time you work with composites such as asphalt shingles, and heat is involved, it is very important not to overheat it or allow it to catch on fire. Composites are also very hard to extinguish once ignited and can burn and get out of control quickly. Proper fire extinguishers should be available in every area where people work, and they should be able to put out various types of fires. If you don't have fire protection, please do not attempt any of the projects in this book that require heat. And always use the appropriate safety equipment, such as goggles, earplugs, and dust masks. Do not wear gloves when working with most tools. Always wear eye protection when working with metal. Do not wear sandals, open-toed shoes, or canvas shoes when working with tools. Avoid loose-fitting clothes that might become entangled in a power tool. Remove rings and other jewelry. Read the owner's manual before using any tool. Never use a tool unless trained to do so. Inspect the tool before each use, and replace or repair it if parts are worn or damaged.

1. Take your 1- × 4-inch × 4-foot board and crosscut it into two 6-inch (15-centimeter) pieces and three 16-inch (40-centimeter) pieces.
2. Take the three 16-inch (15-centimeter) pieces and screw them together with the bottom edges flush with the outside edges of another 8-inch board, making a U shape.
3. Now take the two 6-inch (15-centimeter) pieces and screw them to the ends of the U shape, making a box that's now your block mold. Repeat steps 1 through 3 if you want to make more blocks.
4. Take the mold and coat it with a thin layer of vegetable oil to use as a mold release.
5. Find a hard surface (such as a cement floor) on which to lay the shingles to cut them up. Use your razor knife to cut them into sizes that can easily fit into your pot.

TIP **Cut the shingles with the mineral side down so as to extend the life of your blade.**

6. When you have cut enough shingles to half fill the pot, you are ready to start the melting phase.
7. Place the pot on an outdoor burner in the open air, and add enough vegetable oil to fill the bottom of the pan to where it looks just a little more than wet (Figure 5-41).
8. While wearing gloves, turn on the heat, and add just small pieces of shingle to the layer of oil as it warms. Use the stick or utensil to move the pieces of shingle around in the heating pan of oil (Figure 5-42).
9. As soon as the shingle softens and begins to melt (turn the heat down if you see or smell smoke), start adding more of the shingles while stirring, and keep adding shingles until you have enough to fill the mold (Figure 5-43).

Figure 5-41

Figure 5-42

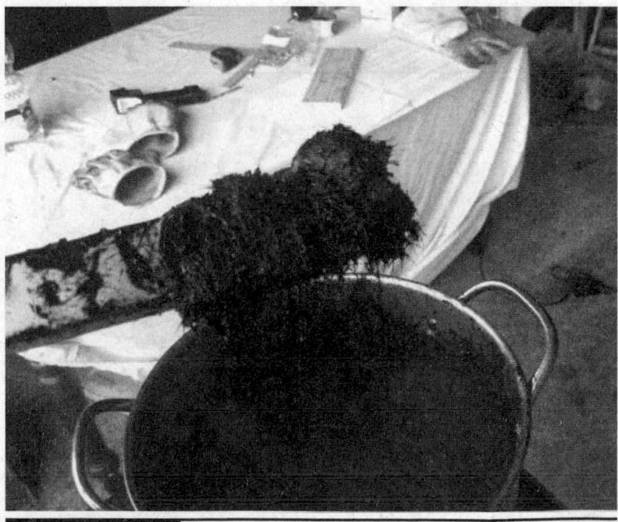

Figure 5-43

10. Turn off your burner, and carefully, with your gloves on and using your stick/utensil, pour the softened asphalt into the mold, and fill any cavities you may see (Figures 5-44 and 5-45).
11. Let the block cool, and then turn the mold upside down, tapping the mold opening on a solid surface to release the block from the mold (Figures 5-46, 5-47, and 5-48).
12. Repeat the preceding steps to make as many blocks as you want.

Figure 5-46

Figure 5-44

Figure 5-47

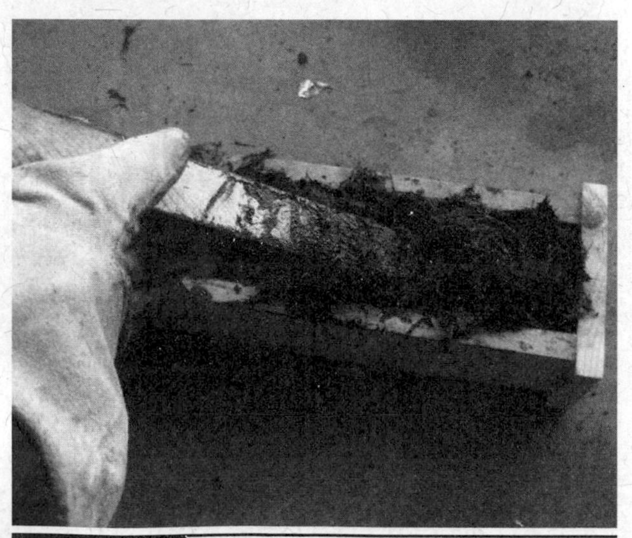

Figure 5-45

Figure 5-48

Project 6
Make a Road-Patch Compound from Recycled Asphalt Roofing Shingles

The use of recycled asphalt roofing shingles to make road patch is a practice that has been used for years. Road-patch compounds made with recycled asphalt roofing shingles have a longer life than other patch materials. This is true because the fibers from the felts or the fiberglass in the recycled shingles add durability. The patch material is easy to apply. A pothole is simply filled approximately an inch over its opening, and the patch is compressed by vehicle traffic.

WHAT YOU'LL NEED

- A small pile of spent asphalt shingles
- A few shovelfuls of small pea gravel or stone
- A small amount of vegetable oil
- A good-sized stick or old utensil to mix, stir, and scrape out the pot
- A electric hot plate or gas burner that you can use outside
- A old 8-quart or larger pan with a handle(s)
- A 4- × 4-inch waist-high wooden post
- A pair of heat-resistant gloves
- Safety goggles
- A razor knife
- A shovel
- A broom (Figure 5-49)

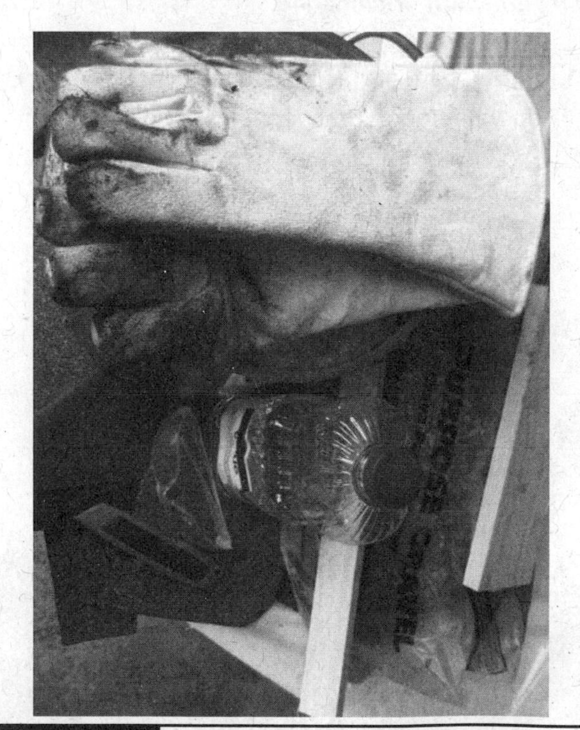

Figure 5-49

Potholes 101

The asphalt used in roads and driveways consists of a rigid mixture of tar and gravel. Asphalts shrink in the cold and expand in heat. Asphalt actually can flow "slowly" when force is placed on it. The elastic nature of "new" asphalt helps it to withstand the shrinking and expanding caused by temperature changes and the weight of traffic on it. However, as asphalt ages, it becomes rigid and brittle because the once-liquid portions oxidize slowly, making it easy to crack, break, and crumble. Oxidized asphalt can no longer stretch to accommodate road stress and performs more like a solid piece of material that loosens from the surrounding asphalt. When traffic presses on it, it loosens further and can be removed as one large piece or hundreds of smaller ones. Water also can increase the hydraulic effect of splashing as water is pushed into the cracks over and over again. In northern climates, the water freezes and expands, making even more cracks.

Fixing a pothole is as simple as cleaning away the oxidized, cracked, and loosened asphalt pieces; drying the pothole; and filling it with fresh asphalt.

Let's Start

TIP Any time you work with composites such as asphalt shingles, and heat is involved, it is very important not to overheat it or allow it to catch on fire. Composites are also are very hard to extinguish once ignited and can burn and get out of control quickly. Proper fire extinguishers should be available in every area where people work, and they should be able to put out various types of fires. If you don't have fire protection, please do not attempt any of the projects in this book that require heat. And always use the appropriate safety equipment, such as goggles, earplugs, and dust masks. Do not wear gloves when working with most tools. Always wear eye protection when working with metal. Do not wear sandals, open-toed shoes, or canvas shoes when working with tools. Avoid loose-fitting clothes that might become entangled in a power tool. Remove rings and other jewelry. Read the owner's manual before using any tool. Never use a tool unless trained to do so. Inspect the tool before each use, and replace or repair it if parts are worn or damaged.

1. Find a hard surface (such as a cement floor) on which to cut the shingles. Use your razor knife to cut them into sizes that can easily fit into your pot. Wear eye protection and your heat-resistant gloves (Figure 5-50).

Figure 5-50

2. Place your pot on an outdoor burner in the open air.
3. Fill it halfway with the gravel or rock, and add enough vegetable oil to make the contents look wet (Figure 5-51).
4. With your gloves on, turn on the heat, and add just a few small pieces of the cut asphalt shingles to the gravel and oil mix (Figure 5-52).
5. As the mixture heats, use the stick or utensil to move the pieces of asphalt around in the pot.

Figure 5-51

Figure 5-52

6. As soon as the asphalt softens and begins to melt (turn the heat down if you see or smell smoke), start adding more of the shingles while stirring, and keep adding them until you have enough to completely cover all the gravel with molten asphalt (Figure 5-53).
7. Turn off the burner, and with your gloves on, carry the pot of hot asphalt to your clean and dry pothole, and fill the pothole (Figure 5-54).
8. Repeat the preceding steps to fix as many potholes as you want.

Figure 5-53

Figure 5-54

How to Use Your Road-Patch Compound

1. It's best to fix a pothole on a warm, dry day. Use a shovel to dig out any loose asphalt or gravel in the hole itself. Be sure to get rid of loose asphalt materials and dirt. You'll need to get right down to the compacted gravel base of your driveway if you can.
2. After you've removed the loose pieces, use your broom to clean up the hole. Make sure that you get rid of all the loose gravel in the hole, or your repair won't bond well.
3. Fill your newly cleaned-up and dampened hole with asphalt cold patch to about 1 inch above the surface. Starting at the outside, tamp the patch down using the end of your waist-high 4- × 4-inch post. If you have a tough time getting your hands around the post, you can buy a hand tool specially designed for tamping or rent a power tamper at a tool rental store. An alternative is to lay a piece of cardboard on the patch and drive your car over it a few times (Figures 5-55 and 5-56).
4. Finally, sweep some dust over your patch so that it will blend in with your driveway and the patch material won't stick to your shoes or tires (Figure 5-57).

Figure 5-55

Figure 5-56

Figure 5-57

CHAPTER 6

Recycling Paper and Cardboard Fiber Projects

Paper is a product that dates back many hundreds, if not thousands, of years. The word *paper* is derived from the name of the reed plant *papyrus*, which grew densely along waterways in ancient Egypt. Papyrus is made from the fibers found in the flower stem of the papyrus plant. Sliced pieces were pressed together, dried, and then used for writing or drawing. Today's paper is still made from plant fibers called *cellulose* found in wood, cotton, and flax. The most general uses for paper are for writing, printing, cardboard boxes, disposable dinnerware, grocery bags, and various craft supplies. Although paper can be made from almost any plant, most modern paper is made from trees and recycled paper. Higher-end and industrial papers (e.g., paper money) come from other common plants used for making paper such as flax (linen) and cotton.

Farm-Fresh Paper

The process for making wood into paper has changed many times over the years. Paper production today is done strictly by mechanical means, but the basics remain the same. Production most often begins with small trees and leftover scraps from lumber mills. Today, managed resources (i.e., trees) used for manufacturing paper are often grown on tree farms. Some farm trees are planted and managed solely for paper production. Each tree is replaced as it is harvested to keep the supply steady and plentiful.

Two types of wood are used in large-scale papermaking. Hardwood trees produce a smoother, shiny paper surface, but their smaller fibers tend to make the paper weak. Softwood trees produce a stronger paper, but the finished product is rough and flexible, making it no good for writing or printing. Paper today is made from a mix of the two types of wood, resulting in a strong and smooth paper. Paper used for grocery bags, crafts, and shipping materials is almost always made of softwood tree fibers.

After the trees are harvested for paper production, they are hauled to the paper mill to be cleaned and stripped of bark. The wood then is chopped into easy-to-manage wood chips. If recycled paper is going to be used, it's added to the wood chips at this point in the process.

Pulp Friction

The next step is turning the wood chips into a semiliquid pulp. Some paper mills have machines that grind the wood, and some use a pulp digester. The digester uses steam and chemicals to break down the wood. The end result is a pulp consisting of water, wood fibers, wood resins, and lignin that are separated from each other. The pulp is further processed, leaving only water and wood fibers. The pulp is bleached at this point in the process if needed.

When recycled paper is turned back into paper, the recovered paper is cut into small pieces. Heating the mixture breaks the paper down more quickly into tiny strands of cellulose (organic plant fiber). Eventually, the old paper turns into a mushy mixture also called *pulp*. The recycled-content pulp is forced through screens containing holes and slots of various shapes and sizes. The screens remove small contaminants such as bits of plastic and globs of glue. This process is called *screening*. The result is clean pulp that is ready to be made into paper. The recycled-fiber pulp can be used alone or blended with new wood fiber to give it extra strength or smoothness.

Paper Cuts

The pulp mixture is sprayed onto a wire belt. The belt lets water seeps through, leaving a mat of pulp. The remaining fibers begin to bond to one another, and the mat then is fed through rollers that extract even more of the water.

The pulp mat then is run through large steam-heated rollers. Heat and pressure are used to dry the remaining water from the pulp mat, turning it into paper. The pulp mat may be run through any number of these rollers until the desired result is achieved. Sometimes as many as a dozen or more trips through the rollers are needed.

Paper is often recycled back into paper and paperboard products. Recovered paper frequently is recycled into products of similar or lower grade than the original paper product. For example, old corrugated boxes are used to make new recycled corrugated boxes, and recovered printing and writing paper can be used to make recycled office or copy paper.

Recovered paper can be used in a variety of other products as well. Recycled-paper pulp can be molded into egg cartons and fruit trays. Recovered-paper pulp can be used for fuel, ceiling and wall insulation, roofing felt, and even animal bedding. In my hometown, cat litter is made from recovered newspaper.

Types of Paper Found in Our Homes

Here is a list of the types of paper and paper products found in my home as well as yours. All these papers are acceptable material feedstocks for the projects in this chapter.

Abrasive paper. Sandpaper is paper covered on one or both sides with abrasive mineral or sand powder.

Absorbent paper. Paper towels and paper napkins have the specific characteristic of absorbing liquids such as water. These papers are soft, loosely felted, and fluffy.

Acid-free paper. Acid-free paper does not contain any acidic substance that may affect the acid-sensitive materials often used in crafts, scrap books, and photo albums. Acid-free paper also has antirust properties and is used for wrapping metal products that are being shipped to help protect them.

Air filter paper. This type of paper is used in car air filters, some furnace filters, and vacuum bags. It is a fibrous type of paper that removes suspended particles such as dust.

Album paper. This paper is used in photographic albums and is also acid-free. It has a soft surface that will not wrinkle or warp when photographs are pasted or glued on it, and when wet with adhesive, the adhesive will not bleed through easily.

Alligator imitation paper. This paper is used for book covers, wallets, and large envelopes. It is textured to resemble alligator skin/leather.

Aluminum paper. This is a wrapping paper made by mixing aluminum powder into the pulp or by coating the pulp mat with aluminum powder. It is used in food and candy wrapping.

Announcement-card paper. This is a high-quality paper with matching envelopes that generally is used for stationery, announcements, wedding invitations, and greeting cards.

Antique paper. This is a printing paper that has bulk and is distressed with a rough or mat surface.

Art paper. This is a high-quality heavily coated printing paper with a smooth surface. Art paper requires an even, well-closed surface for predictable ink absorption.

Bacon paper. This is a thick, almost plastic-like paper that is grease-resistant, usually made in a paper process called glassing. It is used to wrap bacon or other fatty meats from butcher or deli shops.

Bag paper. This is any paper to be used in the manufacture of paper bags.

Baker's wrap. This is the thin paper used by bakers for wrapping bakery goods.

Banknote or currency paper. This type of paper is used for currency and has a very high folding endurance, permanency, and tensile strength. It is also suitable for color printing, watermarking, and other falsification safeguards such as an embedded metal strip.

Barrier paper or wrap. This is treated, coated, or laminated paper that offers resistance to the passage of vapor, gas, moisture, oil, water, or other fluids.

Black waterproof paper. Also called *tar paper*, this is an asphalt-impregnated paper used as a moisture barrier for roofs, walls, and floors in buildings and homes.

Blood-proof paper or butcher paper. This is a high-strength paper that has resistance to animal blood. It is used for wrapping fresh meat. It is coated with a wax emulsion or other liquid barrier material.

Board. This is thick and stiff paper, often consisting of several plies, that is used widely for packaging or box-making purposes.

Bogus paper. This is made from recycled fiber or inferior pulps to imitate higher-quality paper grades. Gray bogus paper is used for insulation, packaging material, bedding, and a variety of other industrial and agricultural purposes.

Business-form paper. This is paper used for business forms and data processing, such as computer printouts.

Burned or weathered paper. This is paper that has been discolored and is brittle but otherwise intact.

Candy-twisting tissue. This is a lightweight paper that is generally waxed for wrapping candy taffy.

Carbon paper. This is a thin paper coated in a waxy ink that is used to produce carbon copies in receipt books.

Carbon-less carbon paper. This is a paper that uses a chemical reaction between contacting coatings to transfer an image when pressure is applied.

Cardboard. This is a thin, stiff paperboard made of pressed paper.

Catalog or Yellow Pages paper. This is a lightweight and good-strength paper typically used for catalogs and telephone directories.

Check paper. This is a strong, durable paper made for the printing of bank checks. This paper is designed to react with a wide range of ink solvents. It gives a characteristic colored stain on contact with acid, alkali, bleach, and organic solvents such as acetone and ethanol.

Cheese-wrapping paper. This is the type of paper used by the food industry to wrap cheese or to keep slices of it from sticking together.

Chipboard. This is a paperboard that is thicker than cardboard, and it is used for backing sheets on writing paper, partitions within boxes, soda can boxes, and shoe boxes.

Cigarette rolling paper. This is a lightweight paper that has calcium carbonate as filler to control the burning rate and improve glowing. It also has very long fibers such as jute or cotton to achieve high strength.

Coffee filter paper. This is used for coffee filtering. It has no impurities and great wet strength and is able to withstand boiling water.

Copier or laser paper. This includes lightweight grades of quality and dimensionally graded paper used for copying correspondence and documents.

Correspondence paper. This is any writing paper, and it is available in attractive finishes, weights, and colors.

Corrugated board. When this corrugated board is glued to another flat sheet of board, it becomes single-faced corrugated board; when glued on both sides, it becomes double-faced corrugated board or corrugated (shipping) container board.

Corrugated medium. This is the wavy center of the wall of corrugated board, and it is designed to cushion a product from shock during shipment and to add strength to the packaging. It often contains up to 100 percent postconsumer recycled fiber content without reducing its ability to protect the product.

Cotton or rag paper. This is paper made with a minimum of 25 percent cotton fiber. It is often found as paper towels in machine and automotive repair shops.

Crepe paper. This is a lightweight paper that normally is colored; it is used for party decorations.

Damask paper. This is a paper with a finish that resembles linen; it is used for formal printing such as work résumés.

Directory paper. This is a lightweight grade of catalog or printing paper with good strength, high opacity, and good print ability; it is used frequently for printing telephone directories.

Document paper. Document paper is paper with a high aging resistance that is used for documents that have to be preserved for a long time.

Drawing paper. This is a flat-finished paper that is of good quality and can withstand friction from erasing.

Envelop paper. This is paper made specifically for cutting and folding of envelopes in high-speed envelop machines.

Fluorescent paper. This is paper coated or surface-treated with fluorescent dye to make it glow in the dark or under a black light.

Folding box board. This is a paper used to make consumer packaging cartons such as cereal boxes.

Fruit wrapping paper. This is a lightweight tissue used for wrapping fruit for shipment.

Index paper. This is a stiff, inexpensive paper with a smooth finish that makes it a preferred choice for business cards.

Ledger paper. This is a strong paper usually made for accounting and records. It has good erasure and pen writing characteristics.

Litmus paper. This is a paper saturated with a water-soluble dye extracted from certain lichens (litmus). The resulting piece of paper becomes a pH indicator that is used to test materials for acidity. Blue litmus paper turns red under acidic conditions, and red litmus paper turns blue under basic conditions.

Magazine paper. This is any paper designed to be used for printing magazine, books, or junk mail.

Matte-finished paper. This is a dull-finished paper.

Mulberry paper. Mulberry paper has distinct fibers running through it and looks handmade. It is very attractive and can be used in all sorts of crafts applications.

Natural-colored or self-colored papers. These are papers in which color is obtained only from the wood fibers used to make it. No dye or pigment is added.

Newsprint. This is paper manufactured mostly from mechanical pulps specifically for the printing of newspapers.

Playing-card stock. This is a card stock made stiff by pasting thin sheets of paper together, making a board. Then it is given a coating that will take a high polish.

Postconsumer waste paper. This consists of paper materials recovered after being used by consumers.

Poster paper. Poster paper is often large, mostly colored paper that has been made weather-resistant.

Preconsumer waste paper. This is paper recovered after the papermaking process but before being used by consumers.

Rag paper. Rag paper is made mostly from fibers consisting of cellulose, such as cotton, linen, and hemp. Rag papers and rag-containing papers are used for bank notes, deeds, documents, books of account, maps, and elegant writing papers.

Recovered paper. This is paper recovered for recycling into new paper products. Recovered paper can be collected from industrial sources or from household collections.

Sack paper. This is brown bag paper; it is a high-strength kraft paper used for sacks.

Sanitary papers. These paper grades are used to make toilet paper and numerous other sanitary products such as facial tissue, kitchen wipes, paper towels, and cosmetic tissues.

Security paper. This is paper that includes identification features such as metallic strips and watermarks to assist in detecting fraud and to prevent counterfeiting.

Self-adhesive paper. This paper is used essentially for labeling purposes. It has a self-adhesive coating on one side and a surface suitable for printing on the other. The adhesive is protected by a laminate that enables the sheet to be fed through printers or printing machines. The laminate is stripped off when the label is ready to be applied.

Tea bag paper. This paper is used to hold tea leaves. It has high liquid permeability and withstands boiling water.

Thermal paper. This is paper with a heat-sensitive coating on which an image can be produced by the application of heat.

Thin paper. This includes carbonizing, cigarette, bible, air mail, and similar papers.

Tissue paper. This is a lightweight thin sheet soft-feeling paper. Its uses in our homes include toilet paper and facial tissue, napkins and paper towels, and other special sanitary papers. Commercial and industrial tissues find use in hospitals, restaurants, businesses, and institutions. Some tissue papers are decorative papers that are glazed or unglazed or include crepe and are used in wrapping gifts or in decorating.

Velvet-finish paper. This is paper with a smooth finish that simulates velvet.

Wadding. This is a loosely matted fiber pad used in packaging, thermal insulation, and acoustical applications. It is also used in diapers and as absorbent material in other sanitary products.

Wallpaper. This is a paper used for wall covering.

Waxed paper. This is paper that is impregnated with paraffin wax.

Wet wipes. This is folded absorbent tissue used for cleaning purposes.

The Paper Trail

Our projects in this chapter involve a process that is very similar to how paper is recovered and recycled today by manufacturers. We are going to first partially pulp the paper with water and combine this pulp with a mineral binder so that we can cast it into usable items for our homes. We don't need to be concerned about inks and contaminants, and we also can recycle a larger number of paper types than the professionals do.

Project 7
Make Concrete Papier-Mâché

Concrete papier-mâché, also known as *papercrete*, is a alternative building material consisting of a mixture of portland cement and recycled paper fiber (slump) that, when cured, makes a lightweight concrete product. Strength is obtained when the paper fiber is thoroughly coated by the portland cement.

Paper fibers are porous and have lots of surface area. This allows for supersaturation when paper is immersed in water. The long paper fibers experience a stacking effect, providing a body matrix that can support the cement, which then crystallizes in a three-dimensional matrix. Papercrete was patented over some 80 years ago and is an environmentally-friendly material owing to the significant recycled content. A number of people have patented several different methods and recipes to make papercrete more marketable as a commercial building material, as well as ways to mass produce it into products. The basic approach to this project is to pulp paper fibers, mix them with concrete, and cast them into a dimensional product.

Mass Production

To mass produce concrete papier-mâché or papercrete, I would first shred the recycled paper and then mix it with water in a large tank or vat to create a uniform pulp. In a second mixer, I would add the portland cement to the paper pulp until the proper consistency is met. I then would extrude the mixture into measured ingots via a profiled die and a moving belt that feeds another set of dies that press or cast those ingots into products that are then allowed to cure and dry.

Safety First

Anyone handling concrete for a major project or just a home patio, sidewalk, or concrete papier-mâché should understand and practice a number of basic safety tips concerning protection, prevention, and commonsense precautions. The tips that follow concern protecting your head, back, skin, and eyes.

- Be careful how you move heavy materials such as portland cement, paper, sand, and water. This can be very strenuous to the average person's back. Most of these materials are heavy even in small quantities.
- Take care to lift materials properly by keeping your back straight and your legs bent to avoid serious back strain.
- Lift heavy materials properly by keeping them waist high and centered between your legs to lessen the chance for injury.
- Ask a coworker or a neighbor for help.
- Use mechanical or manual equipment whenever possible.
- Do not lift wet concrete. Use a wheelbarrow or push the concrete with a shovel or similar tool.
- Watch for skin irritation and chemical burns when working with fresh concrete. Chemical

burns can result from ongoing contact between fresh concrete and skin surfaces, eyes, and clothing.

- Wear protective clothing such as waterproof gloves, long-sleeved shirts, and long pants, and keep the concrete from making contact with your skin.
- Use waterproof pads to protect your skin, knees, elbows, and hands from contact with fresh concrete.
- Avoid direct skin contact with sand and aggregate, both of which are very abrasive. Wet cement is caustic and injurious to your skin.
- Try not to handle portland cement directly because it will draw moisture from your skin and cause it to crack and bleed.
- Wear clean, dry clothes, and try to stay dry so as to not transfer the alkaline by means of hygroscopic effects to the skin. Rinse clothing saturated with wet concrete quickly with fresh water.
- Wash away potential hazards, and take a bath or a shower at the conclusion of your project.

First Aid for Contact Chemical Burns

Immediately flush eyes and skin with clean water. If minor burns persist, see a physician. Seek immediate medical help if burns affect a large area of your skin or appear to be deep.

Protect Your Eyes

Always wear proper eye protection to protect yourself from splattering concrete and blowing dust that can easily enter eyes during concrete placement. Full-cover goggles or safety glasses with side shields may be necessary depending on the project.

WHAT YOU'LL NEED

- Any type of paper listed in this chapter, shredded
- A couple large bags of portland cement
- A little vegetable oil
- A source of water
- A large plastic tub, wheelbarrow, or cement mixer
- A hard-toothed rake
- A 55-gallon drum
- A shovel
- A hoe (Figure 6-1)

Figure 6-1

Let's Start

1. Fill the 55-gallon drum about two-thirds full with your shredded paper, and add water until the drum is nearly full. Soak the paper overnight or over a few days to soften the fibers and make mixing and pulping easier.

TIP Use graywater from your home if you can. If you don't have a graywater catch system in your home, after you bathe, use a bucket to scoop out the water for this project or for watering plants.

TIP If you can let the paper sit in the water for several days in a dark place at room temperature, microbes will help to break down some of the paper's bonds and make it easier to mix and pulp.

2. Mix 80 percent paper pulp and 20 percent cement in the large plastic tub, wheelbarrow, or cement mixer. This process is one of trial and error. Feel free to customize the amounts you use to suit your own needs. Blend the ingredients until you have a thick consistent *slump* (concrete and paper mix). If you are blending in other ingredients, such as dyes, additives, or modifiers, add them to the slump and mix (Figures 6-2 and 6-3).

TIP You want your slump (concrete and paper mix) to be as firm as possible but yet able to be poured, similar to how soft-serve ice cream would act.

TIP Adding sand will make the finished papercrete heavier, and the block won't shrink as much as it dries.

3. Pour the mix into the form you created for the item or project you want.

4. Allow the mixture to sit in the form until set. Larger forms will require more time to set. Leaving the forms overnight, regardless of size, will ensure that everything has set completely.

TIP Old shoe boxes and thick cardboard boxes make good forms, as well as other regularly shaped containers. The shape you use for your form is your choice and depends on your needs.

TIP Use vegetable oil to coat your form or mold to prevent the dry concrete papier-mâché from sticking to your forms or molds.

5. Remove the finished item from the form and mold.

TIP If you need to cut the cured concrete papier-mâché, just use a saw with a crosscutting blade made for cutting wood.

TIP If you need the concrete papier-mâché to be resistant to water, look for an earth-friendly concrete sealer to do the job.

Figure 6-2

Figure 6-3

Waterproofing

Concrete is porous and will absorb water, which shortens its life span and can contribute to other problems. Many concrete sealers are just surface treatments that don't address moisture deep within the concrete. The best concrete sealers are those that permeate deep and set up hard under the surface and inside the pores of the concrete. You want to use a sealer that is a nontoxic clear solution. It should have no volatile organic compounds (VOCs) and should be brushed or sprayed onto the surface easily. A good sealer has a chemical reaction with the alkali that exists in all concrete. This chemical reaction causes a gel to form, which then hardens to glasslike crystals and fills all the pores within the concrete substrate. This silicate crystalline material remains permanently imbedded in the pores, making the concrete so dense that there is nowhere for water to absorb. This permanent internal seal and water barrier also hardens and dustproofs the surface while increasing its compressive and flexural strength.

Project 8
Make and Use Landscape Block Molds

The landscape blocks made from the concrete papier-mâché can be cut easily with a power saw, so we won't need them to be tapered like conventional landscape blocks. The blocks also can be nailed, glued, or screwed together, so there won't be any need to have a lip on the block. You also can use empty tin or pop cans in your castings to take up space and save materials.

WHAT YOU'LL NEED

- Several empty tin or pop cans
- A small amount of vegetable oil
- A 1- × 8-in × 8-foot board
- 1¼-inch decking screws
- A Phillips drive for the decking screws
- A drill or cordless screwdriver
- A table saw or a miter saw
- A tape measure
- Safety goggles
- A razor knife
- A pencil or pen
- A hammer
- Gloves (Figure 6-4)

Figure 6-4

Let's Start

TIP **Use appropriate safety equipment, such as goggles, earplugs, and dust masks. Do not wear gloves when working with most tools. Always wear eye protection. Do not wear sandals, open-toed shoes, or canvas shoes when working with tools. Avoid loose-fitting clothes that might become entangled in a power tool. Remove rings and other jewelry. Read the owner's manual before using any tool. Never use a tool unless trained to do so. Inspect the tool before each use, and replace or repair it if parts are worn or damaged.**

1. Take the 1- × 8-inch × 8-foot board and cut it into six 16-inch (40-centimeter) pieces (Figure 6-5).
2. Take three of the 16-inch (40-centimeter) pieces and screw them together with the bottom edges flush with the outside edges of another 16-inch (40-centimeter) board, making a U shape (Figure 6-6).
3. Now take two 16-inch (40-centimeter) pieces and screw them to the ends of the U shape, making a box that's now your brick mold. Repeat steps 1 through 3 if you want to make more molds (Figure 6-7).

Figure 6-5

Figure 6-6

Figure 6-7

4. Coat the inside of your mold with vegetable oil to act as a mold release, and fill the mold (Figure 6-8).

How to Use the Mold

1. Place two or three empty cans open side down in the vegetable oil–coated mold. Leave a few inches of space between the cans and away from the sides of the mold to reduce weak spots (Figure 6-9).

Figure 6-8

Figure 6-9

2. Pour your papier-mâché concrete slump to just a little over full into the mold. Tap on the mold with a hammer to help settle the slump.
3. Now use the remaining 16-inch (40-centimeter) board to scrape off, or screed, the remaining concrete from the top of the mold.

Figure 6-10

4. Set the mold aside so that the slump can harden and cure overnight.
5. Once the slump is set up, turn the mold over, and tap the open side on the ground to release the block (Figure 6-10).

Project 9
Build a Concrete Papier-Mâché Landscape Wall or Raised-Bed Garden

Landscape (retaining) walls are used to hold back earthen materials such as dirt that don't stack well and can flow on their own accord or with the addition of water. A landscape retaining wall resists the force that gravity places on the unstable earthen material. Soil can weigh a lot, and it's hard to hold back. Landscape retaining walls need to be built the right way, and this means starting with a good base. The first course of a retaining wall can't just be set on top of the ground. It needs to be placed on a well-compacted base and buried (Figures 6-11 through 6-14).

Figure 6-11

Figure 6-12

Figure 6-13

Figure 6-14

WHAT YOU'LL NEED

- A base material such as gravel
- Landscaping blocks
- Galvanized nails
- Compost
- Dirt
- Erosion-control fabric
- A carpenter's bubble level
- A tamper tool
- A hammer
- A shovel
- A saw

Let's Start

TIP **Use appropriate safety equipment, such as goggles, earplugs, and dust masks. Do not wear gloves when working with most tools. Always wear eye protection. Do not wear sandals, open-toed shoes, or canvas shoes when working with tools. Avoid loose-fitting clothes that might become entangled in a power tool. Remove rings and other jewelry. Read the owner's manual before using any tool. Never use a tool unless trained to do so. Inspect the tool before each use, and replace or repair it if parts are worn or damaged.**

1. Dig out a foundation for your wall that is approximately 2 feet (60 centimeters) wide and 8 inches (20 centimeters) deep.
2. Next, fill the hole with your base material. Gravel is the best base to use, but dirt also can be used.
3. After filling the hole with base material, now you need to compact it to give the blocks a firm base on which to stand. The easiest way is to rent a tamper, or you can do it by hand.
4. After tamping, start laying the block. Set the blocks in a row, and check them with your level. Cut the blocks if you need to. If one side or the front or back is too high, move it with your hammer. Hit the block until it's level. Do the same for each block on the bottom level.
5. After your base course is laid, the rest is easy. Just stack the next row with an offset of one-half block.
6. After the second row of blocks is in place, the erosion-control fabric should be nailed to each block with a few feet of it lying behind the blocks. Repeat this step on every other row as you build the wall. Skip the last three rows at the top so as not to interfere with plant roots.
7. After laying the fabric, backfill on top of it to make a "dead man's anchor." Use the dirt you dug out from the base. Start backfilling as you get to the second or third row (Figures 6-15 through 6-17).

Project 10
Make a Concrete Papier-Mâché Shed or Animal Shelter

Making a shelter from your landscape blocks is very similar to building a retaining wall but with four sides and a roof. Any dog would be happy to call this project home (Figures 6-18 and 6-19).

Figure 6-15

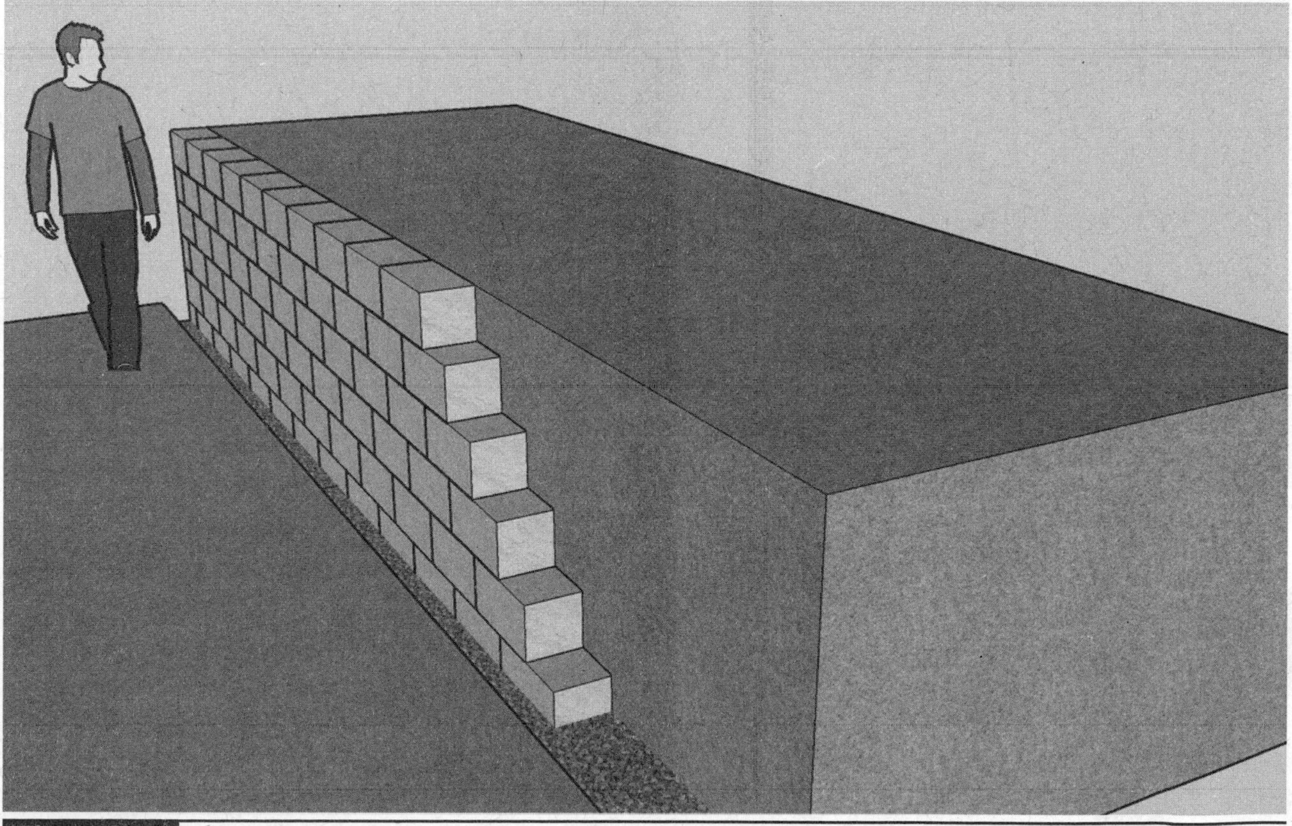

Figure 6-16

Figure 6-17

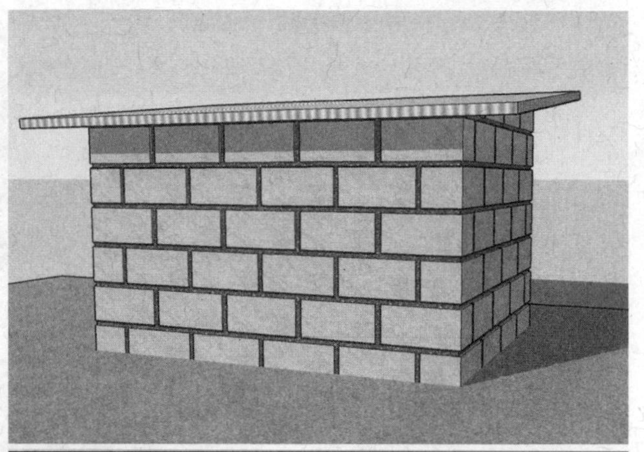

Figure 6-19

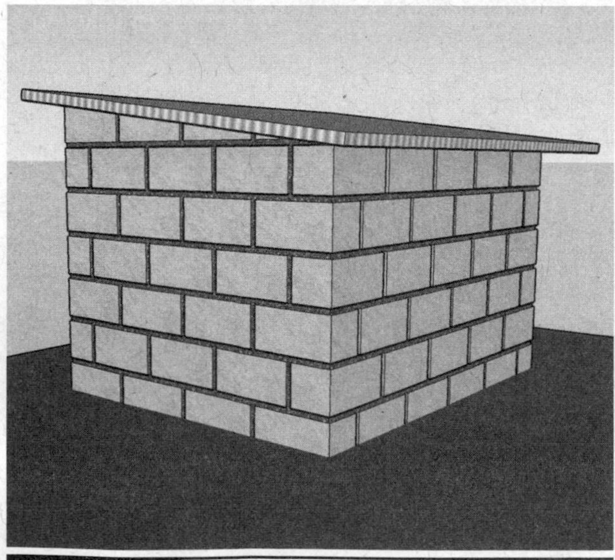

Figure 6-18

WHAT YOU'LL NEED

- Concrete papier-mâché slump
- A base material such as gravel
- Landscaping blocks
- Galvanized nails
- Metal roofing
- Lumber
- A carpenter's bubble level
- A hammer
- A shovel
- A saw

Let's Start

TIP Use appropriate safety equipment, such as goggles, earplugs, and dust masks. Do not wear gloves when working with most tools. Always wear eye protection. Do not wear sandals, open-toed shoes, or canvas shoes when working with tools. Avoid loose-fitting clothes that might become entangled in a power tool. Remove rings and other jewelry. Read the owner's manual before using any tool. Never use a tool unless trained to do so. Inspect the tool before each use, and replace or repair it if parts are worn or damaged.

1. Find a nice level place for your building. Use your shovel to level it if you need to. For a large building project, follow the directions on building a base in the preceding project on building a retaining wall.
2. Pour a 1-inch-deep layer of concrete papier-mâché on your level base. Now start laying the blocks, using your slump as mortar between the surfaces of the blocks. Set the blocks, and check them with your level. Cut the blocks to fit your needs. If one side or the front or back is too high, move it with your hammer. Hit the high side of the block until it's level. Do the same for each block on the bottom level.
3. After your base course is laid, the rest is easy. Just add a layer of concrete papier-mâché, and stack the next row with an offset of one-half block. Repeat this step on every row. Remove excess slump if you want the walls to look smooth.
4. When you reach the desired height, let the slump you used as mortar set up.
5. Cut a piece of metal roofing to fit over the top of your structure with an overhang of at least 12 inches to keep rainwater away from your walls and floor. You can do a flat roof like that in the figures or get creative and do a pitched or lean-to style roof.
6. Attach whatever roof you choose with nails, and fill any cracks with slump.
7. Paint, stain, or seal the building, place some straw or a recycled blanket on the floor, and let your doggie move into its new home.

Project 11
Make a Large Concrete Papier-Mâché Planter

If your garden landscape or patio needs a large feature or centerpiece, you might consider making a concrete papier-mâché planter. Concrete planters are also ideal when you want to plant large shrubs or trees because such planters have the room and strength necessary to hold the larger root systems these plants produce. This project is also an excellent way to create a border around a patio or deck or along a driveway (Figures 6-20 and 6-21).

Figure 6-20

Figure 6-21

WHAT YOU'LL NEED

- Concrete papier-mâché slump
- A earth-friendly concrete sealer
- Any sturdy item whose shape you want to mimic for your planter
- A handheld board to smooth out the concrete papier-mâché
- A large, thick polyethylene sheet
- A carpenter's bubble level
- A shovel
- Gloves

Let's Start

1. Find a sturdy item from which to make a male mold (plug). The item must have only positive, or protruding, points on it. If there are places in the mold where the concrete papier-mâché can flow into and harden, such as a crevice, you may have to destroy the feature or the planter to remove it from the mold (Figure 6-22).
2. Place the item you want to cast in an area that's fairly level and where it won't be in the way. In this way, you won't need to move it, so the planter will have time to harden and cure.

Figure 6-22

3. Cover the item with the polyethylene, and leave a couple feet of material around the base (Figure 6-23).
4. With your gloves on, begin making your planter by adding a layer of concrete papier-mâché at least 2 inches (50 millimeters) thick around the bottom edge of the item, where the polyethylene-covered floor and the item meet (Figure 6-24).
5. Keep adding the concrete papier-mâché over 2 inches (50 millimeters) thick up the entire side of the item you're casting. If you have trouble with the concrete papier-mâché slump not

Figure 6-23

Figure 6-24

being thick enough to support its own weight, add a little more dry concrete mix to it to thicken it. If thickening it still doesn't help you go as high as you would like, let the first part set up before you add more.

6. When you get to the top of the item you are casting, continue to add the concrete papier-mâché slump until it is over 2 inches (50 millimeters) thick, and then make a level area using your carpenter's bubble level. This is what your planter is going to sit on, so make sure that it is wide enough to keep your planter stable (as wide as the planter is high) (Figure 6-25).
7. Before the concrete papier-mâché slump sets up, make sure to punch a few holes with your fingers into what will be your planter's bottom for water to drain out (Figure 6-26).
8. After it cures, flip it over and remove the casting plug and plastic.
9. Then you can stain, paint, or waterproof your planter.
10. Now just add dirt and a plant.

Figure 6-25

Figure 6-26

Project 12
Make a Concrete Papier-Mâché Flower Pot or Decorative Bowl

Flower pots are great gifts, especially if there is a plant or flowers in them. A homemade pot is a wonderful personal gift, and one made from recycled resources is even better. Even if you don't have plants, I'm sure that you know someone who does.

WHAT YOU'LL NEED

- Concrete papier-mâche slump
- An earth-friendly concrete sealer
- Any item to which you want to size your pot, such as an old flower pot, butter tub, or even a toy ball
- A large plastic bag
- A carpenter's bubble level
- A board (Figure 6-27)

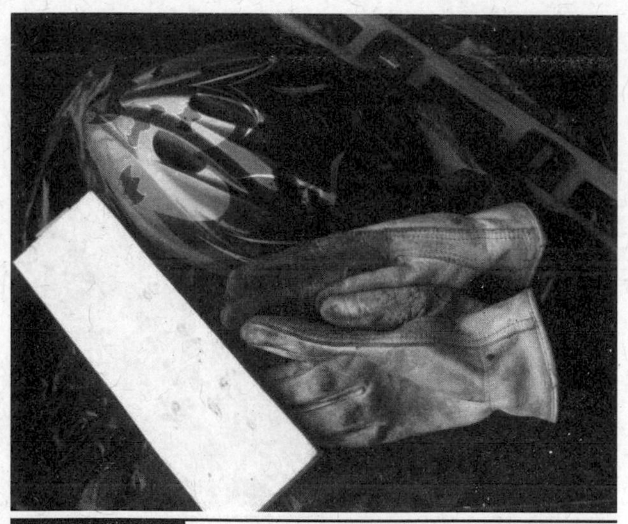

Figure 6-27

Let's Start

1. Place the item sizing your flower pot on a flat surface.
2. Open the plastic bag, and place it over the top of the item you are casting (Figure 6-28).
3. Now add concrete papier-mâche slump to the open bag, and fill it to the point that the item underneath the bag is covered with the slump (Figure 6-29).
4. Carefully level the top of the slump in the bag, and tie the bag closed to keep its shape (Figure 6-30).

Figure 6-28

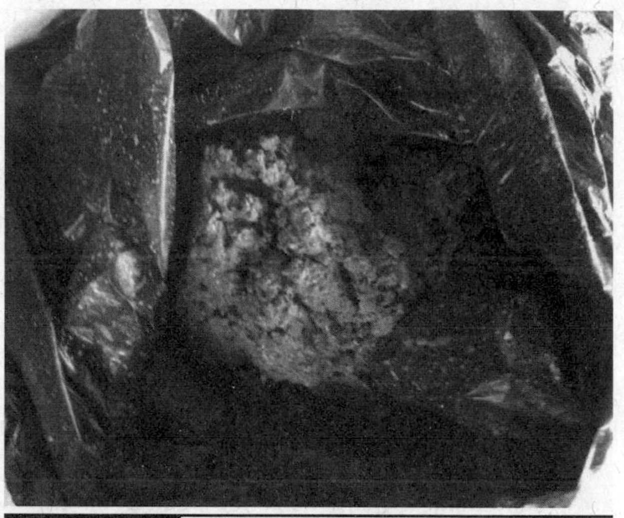

Figure 6-29

Figure 6-30

5. Let it sit to harden and cure.
6. Once it is solid, remove the item from underneath the bag, and then remove the bag from around the pot/bowl you just made.
7. Now you can stain, paint, or waterproof your pot/bowl (Figures 6-31 and 6-32).

TIP **If you waterproof the pot, you may need to drill a hole in the bottom to let excess water drain out.**

Figure 6-31

Figure 6-32

Project 13
Make a Concrete Papier-Mâché Birdbath

Birds, like all animals, need a fresh source of water at all times. A birdbath will give birds a place to drink and bathe and may increase the number of birds that visit your backyard. The size of the birdbath should depend on how many and what type of birds you want to attract. Birdbaths can range from a small dish to a complete garden pond. The smaller and more shallow the birdbath, the smaller and less varied the birds will be.

However, a birdbath can be too deep for some birds. Smaller birds may not be able to use a birdbath with more than an inch of water. Some people find it beneficial to use a birdbath with a shallow edge and a deeper center. The smaller birds can use the outer part of the birdbath, whereas large birds can move toward the center. The most common water depth in a backyard birdbath is around half an index finger deep.

WHAT YOU'LL NEED

- Concrete papier-mâché slump
- An earth-friendly concrete sealer
- Sand or pea gravel
- A plastic sheet
- A carpenter's bubble level

Let's Start

1. Find a flat surface on which to work, and decide how big of a birdbath you want.
2. Once you decided on a size, make a pile of sand or pea gravel on your working surface.
3. Shape the sand or pea gravel to match the interior of the birdbath you want. The depression in the sand or pea gravel is basically going to be the mold (plug) for your birdbath's basin.
4. The depression in the sand or pea gravel should be a little over halfway up your index finger in depth in the center and just to the first joint of your finger at the edge. If you have a hard time keeping the shape in the sand or pea gravel, add a little water to it so that it sticks together like a sand castle does.
5. Carefully cover the mold (plug) you just made with the plastic sheet.

6. Slowly cover the area with concrete papier-mâche slump to a depth of about 2 inches (50 millimeters) all the way to the edge of the mold.
7. Let it sit to harden and cure.
8. Once it is solid, remove the birdbath from the plastic, and clean up your sand or pea gravel.
9. Now you can stain, paint, or waterproof your birdbath.

Location, Location, Location

Place your birdbath on the ground, or make a pedestal if you want. Then just fill the birdbath with clean water and wait for the birds. If you don't see birds in a few days, you may need to move your birdbath. Many birds prefer baths placed out in the open and not too near shrubs or small trees for an escape route. You also can place your birdbath near a bird feeder (not so close that the seeds fall into the birdbath). If predators such as cats are an issue, place the birdbath away from any plants or buildings that can conceal the predator. You want to prevent the predator from having the element of surprise to keep your birds safe. Birdbaths should be cleaned when the water is changed, which should be approximately every two to four days. During the summer months, the water will evaporate more quickly, requiring more refills. Typically, birdbaths can be easily cleaned with a small scrub brush. For heavily soiled birdbaths, you can use a mild detergent for cleaning.

I Thought I Saw a Tweety Bird

Birds are picky when it comes to feeders and housing, and it can be the same for a birdbath. It may be that you live in an area that doesn't attract birds because of a lack of the plants and food sources they prefer. Conversely, you may live in an area that has too many flowers, shrubs, trees, or water sources that birds like, and they may be happy where they are currently feeding and bathing. Another major factor can be your water. If you have left your water out too long, it may have become dirty. Clean your birdbath thoroughly, replacing the water, and then move it to a new location. If you are not using a bird feeder, you may want to consider putting one up to further entice the birds to visit your yard.

Project 14
Make Concrete Papier-Mâché Stepping Stones

Any path around the outside of your home can be made more attractive with stepping stones. On a muddy path, they can be very beneficial by keeping your home mud-free. Stepping stones are a more affordable option than pouring a concrete sidewalk. And when the stepping stones come from recycled items, it makes the path that much sweeter.

WHAT YOU'LL NEED
■ Concrete papier-mâché slump
■ An earth-friendly concrete sealer
■ Plastic shopping bags
■ A large coffee can (Figure 6-33)

Figure 6-33

Let's Start

1. Make sure that your plastic shopping bags are free of holes, and find a dry, flat place to work.
2. Open a shopping bag and put two large coffee can–sized amounts of concrete papier-mâché slump into the shopping bag. Fill the bag where it's going to harden and cure so that you don't risk it tearing and spilling when you move it. Repeat this step for each additional stepping stone you want (Figure 6-34).

Figure 6-34

3. To get a uniform thickness for your stepping stones, try to use the same size bags as well as the same volume of slump for each one.
4. To adjust the thickness of your blocks, just lift the bag slightly and lower it over and over until you reach the desired thickness.
5. Remove the stepping stones from the bags when they have hardened and cured.
6. Now you can stain, paint, or waterproof your stepping stones (Figure 6-35).

Figure 6-35

Project 15
Make a Concrete Papier-Mâché Stepping Stone Walkway

Take a good look around your yard, and find the areas where the grass has been worn down because of frequent travel. This is the best place for a stepping stone path. However, don't feel that you are limited to just those areas. A stepping stone path is also a great way to accent a secluded section of your yard.

WHAT YOU'LL NEED
■ Stepping stones
■ Sand, pea gravel, or recycled rubber mulch
■ A shovel

Let's Start

1. Mark a path for your stepping stones by poking sticks in the ground where you want the path to go. The sticks mark the center of each stepping stone.
2. Remove the sticks one at a time as you create the path. Use the shovel to cut around where you want to place the stepping stone using the stone as a guide, and then remove the stone. Now dig down about an inch more than the actual thickness of the stepping stone. Try to make the bottom of the hole as flat as possible.
3. Place sand, pea gravel, or rubber mulch in the bottom of the hole. This will be your base to help keep the stone from sinking over time. Spread the sand, pea gravel, or rubber mulch in the hole and tamp (pack) it down, making sure that it is flat and level.
4. Place the stepping stone into the hole. Press it into the packed material, and then pack in some of the dirt from the hole around the stone.
5. Repeat these steps for every stepping stone. Make sure that you don't forget to add some rubber mulch, sand, or pea gravel around the stepping stone after you have set it in place (Figure 6-36).

TIP **Setting the stepping stones level with the ground surface prevents a tripping hazard. It also makes it safer to cut the grass without the mower blade possibly striking the stones.**

Figure 6-36

Project 16
Make a Concrete Papier-Mâché Tabletop

Making your own furniture can be a rewarding accomplishment; the item you create also can span generations and add to your legacy. When you use recyclable materials to build your furniture, you also add a legacy of pride and the values you want to pass on to future generations.

WHAT YOU'LL NEED
■ A ¾-inch × 4- × 4-foot sheet of smooth plywood
■ A 5- × 5-foot sheet of 6-mil plastic sheeting
■ Concrete papier-mâché slump
■ A small amount of vegetable oil
■ Two 2- × 4-inch × 8-foot boards
■ 2-inch decking screws
■ Fine sand paper
■ Concrete dye
■ ⅝-inch staples
■ Wire mesh

WHAT YOU'LL NEED

- Plaster of paris
- A Phillips drive for the decking screws
- A drill or cordless screwdriver
- A table saw or miter saw
- Contractor's staple gun
- A tape measure
- Safety goggles
- A putty knife
- A pencil/pen
- A hammer
- Gloves (Figure 6-37)

Figure 6-37

Let's Start

TIP Use appropriate safety equipment, such as goggles, earplugs, and dust masks. Do not wear gloves when working with most tools. Always wear eye protection. Do not wear sandals, open-toed shoes, or canvas shoes when working with tools. Avoid loose-fitting clothes that might become entangled in a power tool. Remove rings and other jewelry. Read the owner's manual before using any tool. Never use a tool unless trained to do so. Inspect the tool before each use, and replace or repair it if parts are worn or damaged.

1. Cut one of the 2-inch (50-millimeter) × 4-inch (100-millimeter) × 8-foot boards into two 4-foot (122-centimeter) pieces. Cut the other 2-inch (50-millimeter) × 4-inch (100-millimeter) T 8-foot board into two 45-inch (114-centimeter) pieces (Figure 6-38).
2. Now screw the two cut 48-inch (122-centimeter) 2-inch (50-millimeter) × 4-inch (100-millimeter) pieces on their edge to the opposite ends of the smoothest side of the 4-foot (122-centimeter) × 4-foot (122-centimeter) piece of plywood.
3. Screw the two 45-inch (114-centimeter) 2-inch (50-millimeter) × 4-inch (100-millimeter) pieces to the open ends of the plywood sheet (making a box), and screw their ends into the two 48-inch (122-centimeter) 2-inch (50-millimeter) × 4-inch (100-millimeter) side pieces. This is your mold (Figure 6-39).
4. Drape the plastic sheet over the open end of the mold, fold the plastic over itself just a little on the edges (so that you're stapling through two layers of plastic), and staple it to the box with a staple every inch or so (Figures 6-40 and 6-41).

Figure 6-38

Figure 6-39

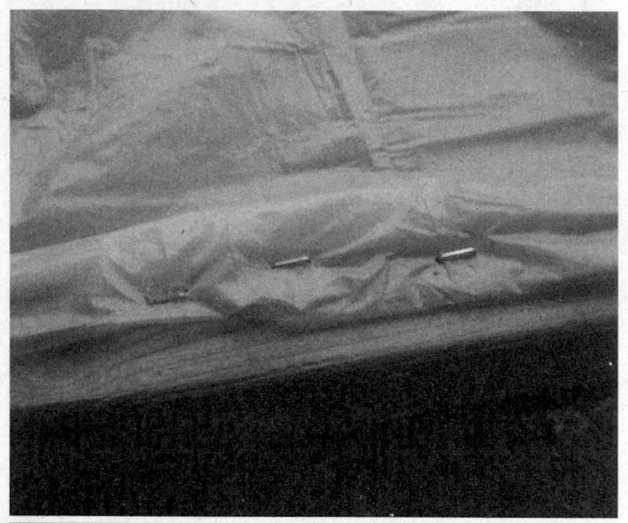

Figure 6-40

5. Moving to the opposite side, stretch the plastic just tight enough to completely touch the bottom of the mold but so that it does not fall into the corners. Now fold the extra plastic over like you did on the first side, and staple it like you did on the first side (Figure 6-42).
6. Repeat the same process on the other sides of the box, making sure that the plastic is just tight enough to completely touch the bottom of the mold but not fall into the corners.
7. Apply a thin layer of vegetable oil to the plastic, and if you want to add some additional color to your tabletop, put a few drops of a concrete dye or stain on top of the plastic with the oil while moving it around randomly to add effect (Figure 6-43).
8. Now add your concrete papier-mâché slump to the mold. Fill it about an inch deep, and tap on all the corners with your hammer to move air bubbles away from the bottom of the mold.
9. Cut the wire mesh to fit in your mold, but be sure to leave a few inches of space between the mesh and the edges of the mold (Figure 6-44).

Figure 6-41

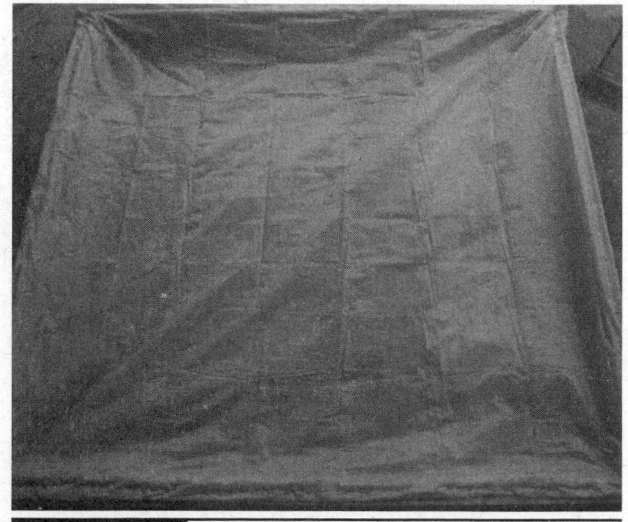

Figure 6-42

Figure 6-43

Figure 6-44

Figure 6-45

Figure 6-46

10. Now cover over the top of the wire with about an inch of the slump, and again tap the edges of the mold with your hammer. Then let it sit to harden and cure.
11. After you have a good hard cure, have someone help you to flip the mold on its side, and carefully remove your tabletop (Figure 6-45).
12. Clean the oil film from your tabletop with some warm soapy water, and wipe it dry.
13. To fill any voids such as air pockets or pinholes, use plaster of paris, and add some some dye to it to blend in with your tabletop (food coloring works with plaster of paris).
14. Apply the plaster to the voids with a putty knife, and let it harden.
15. If you want your tabletop to be smooth, just sand the whole tabletop lightly; if not, just sand the plastered spots to smooth them out.
16. Now you can stain or seal your tabletop. You can use a staining wax if you want for staining and sealing. Or use clear wax for a natural look. If your table is going to be outside, you had better waterseal it.
17. Add your own legs or pedestal, and you have a potential heirloom made by your own two hands (Figure 6-46).

CHAPTER 7

Solar-Powered Compost Projects

THE ORGANIC AND BIODEGRADABLE waste streams from our homes consist of paper, cardboard, food, garden waste, yard clippings, leaves, woody waste, and even animal waste such as cat litter. Through the process of composting, organic wastes can be recycled to produce a natural material that can be used in gardens, landscapes, and flower beds. When mixed with the soil, compost increases the organic matter in soil, which improves its physical properties, supplies essential nutrients, and enhances its ability to support plant growth. Compost also can be applied to the soil surface to conserve moisture, control weeds, reduce erosion, add to the appearance, and keep the soil thermally insulated.

Deterioration and chemical degradation of soil are a severe problem worldwide. It is seen in soil compaction, poor tubes, surface crafting, slow water seepage, slow water drainage, diminished nutrients, low nutrient retention, and decreased crop productivity. This problem is caused mainly by long-term chemical fertilizer application and mechanical tillage. The level of organic matter determines the quality of the soil. All soils have organic content, say, on average, 6 to 15 percent. However, plowing causes the organic matter to decompose quickly, whereas the use of chemical fertilizers discourages the adhesion of organic matter. Year after year, therefore, farmland soils have a lower percentage of organic matter, and the quality diminishes. Thus, currently, most of farmland soils have an organic matter content that is lower than 3 percent.

Compost 101

Gardeners have used compost for as long as there have been gardens. Composting is an efficient method of breaking down organic materials that are beneficial to plants and soil. When materials such as leaves and grass clippings are composted, a microbial process breaks their complex chemical bonds resulting in simpler organic compounds that are beneficial to the quality of the soil. A soil's physical properties, such as infiltration, drainage, and water-holding capacity, usually are improved when composted materials are added.

Requirements for Organic Decomposition

The decomposition of organic materials in a compost pile depends on the activity of the microbes doing the decomposing. The health of these microbes also slows or speeds the composting process. Efficient decomposition occurs when there is adequate air and moisture, when the particles of waste material are small, and when the proper amounts of fertilizer and lime are added.

Aeration

Microbes require oxygen (aerobic conditions) to decompose organic wastes efficiently. Some decomposition will occur in the absence of oxygen (anaerobic conditions), but the anaerobic process is slow and creates foul odors. Because of the odor problem, composting without oxygen is not recommended in residential areas. Turning the pile over (mixing) once or twice a month will provide the necessary oxygen and significantly hasten the composting process. A pile that is not mixed may take three to four times longer to produce useful compost.

Moisture

Moisture is essential for microbial activity. Materials in a dry compost pile will not decompose efficiently. If rainfall is limited, the pile must be watered periodically to maintain a steady decomposition rate. Enough water should be added to completely moisten the pile, but overwatering should be avoided. Excessive moisture can lead to anaerobic conditions, slowing down the process of composting and causing foul septic system–like odors. The pile should be watered enough that it is damp but does not remain soggy. Approximately half its weight in water is a good starting point. The compost is within the right moisture range if a few drops of water can be squeezed from a handful of material. If no water can be squeezed out, the materials are too dry. If water gushes out, the compost is too wet.

Particle Size

The smaller the organic waste, the faster the material will compost and be ready to use. Smaller particle sizes have more surface area for the microbes to react with and are more rapidly broken down by the microbial feast. Some materials should be shredded before they are added to the pile. Moreover, shredding reduces the volume of the compost pile. I rake the the leaves that fall in autumn into small piles, and then I run the lawn mower with a bagger over leaf piles I made.

Fertilizer and Lime

Microbial activity is affected by the ratio of carbon to nitrogen in the organic waste. Microbes require a certain amount of nitrogen to survive and reproduce. Too little nitrogen stalls the composting process, as does having too much carbon. Poultry litter, manure, or blood meal can be used as organic sources of nitrogen. If you can't find a natural source, an off-the-shelf fertilizer with a nitrogen content of 10 percent or higher can be used. Other nutrients, such as phosphorus and potassium, are already present in most of what you compost. During the first stages of decomposition, organic acids are produced, and as the acidity rises, the pH of the compost pile drops. There is no need to add lime to compost unless large quantities of pine waste or vegetable and fruit wastes are composted or the odors are getting out of check.

Materials for Composting

Much of our home waste organic materials is suitable for composting. Yard waste such as leaves, grass clippings, straw, and nonwoody plant trimmings can be composted. Leaves are the dominant organic waste in most home compost piles. Grass clippings also can be composted if need be, but with proper lawn care, clippings do not need to be removed from the lawn, so "cut it high and let it lie." The grass clippings will decay in place and release nutrients, reducing the need for fertilizer. If clippings are composted, they should be mixed with other yard waste because their long, thin shape allows them to compact and restrict airflow in the compost pile. Branches and twigs that are larger in diameter than a pencil should be put through a shredder or chipper first. Kitchen waste such as vegetable scraps, coffee grounds, and eggshells also may be added. Sawdust may be added in moderate amounts if additional nitrogen is applied. Approximately one pound of actual nitrogen is required per hundred pounds of dry sawdust. Wood (fireplace) ashes can serve as a lime source and contribute to higher levels of potassium, but they should be added in only small amounts. Crushed clam or oyster shells, eggshells, and bone meal are alkaline and also can reduce the acidity level of the compost. Regular newspapers can be composted, but be aware that their nitrogen content is low and stalls decomposition. When paper is composted, it should make up no more than 10 percent of the total weight of the material in the compost pile. It is better to recycle newspaper by other means than composting. Other organic materials that can be used to add nutrients to a compost pile include blood and bone meal, livestock manure, nonwoody plants, vegetable and flower garden refuse, fruit and vegetable scraps, hay, straw, and water plants. Animal manure and poultry litter can be added to provide nitrogen.

Materials to Avoid in a Compost Pile

Some materials are a health hazard or can create a nuisance and shouldn't be used to make compost. Human waste should not be used because it can transmit disease. Although animal waste, as well as their remains, such as meat, bones, grease, whole eggs, and dairy products, can be decomposed safely in commercial compostors, these items should be avoided in home compost piles because they may attract pests and create odors. Most disease-causing organisms and weed seeds are destroyed during the composting process because temperatures in the center of the pile reach high enough levels to kill them. Well-shredded citrus rinds (peels), corn cobs, palm leaves, and walnut, pecan, and almond shells also break down.

Location

The compost pile should be located near the place where the compost will be used. Composting is best done in a location screened from your view and that of neighbors. Do not locate the compost pile on a slope that drains to surface water, such as a stream or a pond. Locating the pile near large plants such as trees or bushes is not a good idea because their roots may grow into the bottom of the pile, making handling the compost difficult. The pile will do best where it is protected from wind and is in partial sunlight to help provide heat. The more wind and sun the pile is exposed to, the more moisture it will need.

Preparing the Compost Pile

When a compost pile is started, materials should be added in layers to ensure proper mixing. Organic waste such as leaves, grass clippings, and plant trimmings are put down in a layer 8 inches (20 centimeters) to 10 inches (25 centimeters) deep. Coarser materials will decompose faster if placed in the bottom layer. This layer should be watered until moist but not soggy. A nitrogen source should be placed on top of this layer. Use a finger's depth of livestock manure, and apply a layer of soil or completed compost on top of the manure layer. One reason for adding soil or compost is to ensure that the pile is inoculated with microbes. The use of soil or compost in a new compost pile is optional. This is just one way to ensure that activator microbes are present in the new compost. Adding soil does help to reduce leaching of mineral nutrients such as the potassium released during decomposition. Repeat the sequence of adding organic waste, fertilizer, moisture, and soil or old compost until the pile is completed. If only shredded tree leaves are to be composted, layering is not necessary. Fallen leaves can be added as they are collected. The leaves should be moistened if they are dry. Since dead leaves lack the nitrogen needed for rapid decomposition, add a finger's depth of livestock manure and a layer of soil or completed compost on top of the manure.

Use of Compost to Improve Soil

Compost is used as an organic amendment to improve the physical, chemical, and biological properties of soils. Compost helps the soil to become an ideal environment for healthy plant and root growth. Annual additions of compost to soil will help to create a more desirable soil structure and will make the soil much easier to work with. Remember, compost may enrich the soil, but it also releases nutrients slowly and sometimes does not contain enough nutrients to supply all the needs of growing plants. Play it smart, and conduct soil tests to further understand the needs of your own lawn and garden.

- The compost adds air spaces to the soil, and incorporating it relieves compacted conditions.
- Adding compost increases the moisture-holding capacity of sandy soils, reducing drought damage to plants.
- When added to heavy clay soils, compost may improve drainage and aeration.
- Adding compost increases the ability of soil to hold and release essential nutrients.
- The activity of earthworms and soil microorganisms beneficial to plant growth is also promoted.
- Other benefits include improved seeding and water infiltration as a result of reduced soil crusting.
- Adding compost to soil also can reduce plant disease and root rot. The microbes in compost compete better with disease-causing microbes for the organic nutrients secreted from plant roots and help to keep them inactive.

TIP **To improve the physical properties of your soil, till a finger's depth of well-decomposed compost deep into the soil.**

Charcoal Is Also Beneficial for Soil

Adding charcoal to your compost is a great way to further improve soil. This is an ancient technique discovered by pre-Colombian tribes in the central Amazon region. They were adding charcoal to soil from as far back as 1,500 years ago. The charcoal they made 1,500 years ago was still working even after 1,000 years of crop cultivation without any other fertilization. Charcoal is full of tiny pores, and when charcoal is applied to soil, its pores allow air to diffuse into the soil. Plant roots need the airspaces to grow effectively. The porous charcoal also retains water and nutrients, storing them for growing plants. Charcoal is very stable and will not decompose to carbon dioxide like other types of biomass. Once applied, therefore, charcoal will stay in soil for hundreds to thousands of years. Charcoal comes from raw biomass that has been burned under low-oxygen conditions, which converts it into the charcoal, and at least 50 percent of the biomass's carbon will be permanently fixed into the charcoal, which doesn't biodegrade easily. Thus the major part of the carbon will stay in solid form and won't be released as carbon dioxide to the atmosphere. Charcoal is a good way to lock away carbon in a solid form and keep it out of our air. We could take all the tree leaves, corn stalks, and wood chips and turn them into charcoal and lock away carbon and create a great soil amendment at the same time.

Project 17
Make Your Own Garden-Friendly Charcoal from Biomass Waste

The method we will use here employs the incomplete combustion of organic matter, which is to become charcoal. The rate of combustion is controlled by regulating the amount of oxygen allowed into the burn, and the process is stopped by excluding oxygen before the charcoal itself begins to burn. This is an age-old method used by colliers to make charcoal in a pit, pile (clamp), or more recently, metal or masonry chambers (kilns). Store-bought briquettes have additives in them, so don't use them, and the natural chunk charcoal is a little expensive to be adding to your dirt.

WHAT YOU'LL NEED

- Woody and fibrous biomass waste
- A metal 55-gallon drum
- A piece of sheet metal about 12 inches across
- A saber saw or metal grinder
- An extra pair of hands
- Fire-resistant blocks
- A good bit of sand
- A bucket of water
- A ⅜-inch drill bit
- A drill
- Eye protection
- A shovel
- Gloves
- Matches (Figure 7-1)

Figure 7-1

Let's Start

TIP **Fire is very dangerous, so making your own charcoal is also a dangerous task. If you have any doubts, don't do it. Check with your local authorities for outdoor burning restrictions. During especially dry seasons, even recreational and cooking fires can be restricted. Check the weather forecast. Do not use flammable liquids. Keep first aid supplies and emergency telephone numbers accessible at all times. Avoid areas with overhanging branches, steep slopes, or dry grasses. Clear the area of all debris, down to bare soil. Have a bucket of water and shovel nearby to put out the fire if needed. Never leave a fire unattended. Burn only natural vegetation such as wood. Wear eye protection, and have someone help you with any heavy lifting.**

1. Cut a hole about 6 inches (15 centimeters) in diameter in the center of the bottom of your metal 55-gallon (208-liter) drum (a plastic drum would melt and catch on fire). Make sure that your piece of sheet metal fits over the 6-inch (15-centimeter) hole completely because you're going to use it to cover the hole to dampen the combustion (Figure 7-2).
2. Flip the drum over, and add small and narrow pieces of dry woody and fibrous biomass waste to the drum.

Figure 7-2

3. Once the drum is full, have someone help you tip it over. If stuff falls out, you can stuff it back in through the 6-inch (15-centimeter) hole you just cut.
4. Now, using three or four fire-resistant blocks, tip the drum and have your helper place blocks under the edges of the drum to keep the drum lifted off the ground (Figure 7-3).
5. Add some kindling to the airspace below the drum to help you light your fire later.
6. Put sand around the drum but not touching it (yet). Leave enough space around the drum so that you can light it. You want enough sand around the drum so that it reaches about 6 inches (15 centimeters) high on the drum and 6 inches around it. The sand is used to restrict the airflow from the bottom so that you can make the charcoal (Figure 7-4).
7. Pour a little water directly on the top of the drum so that you can see it boil away to give you an idea when to dampen the flame.
8. Now ignite the biomass from the bottom, and let the flames spread. You should see a lot of smoke (mostly steam/water vapor) coming out of the 6-inch (15-centimeter) hole on the top.
9. Let the fire burn while watching for water you placed on the top surface of the drum to boil (Figure 7-5).
10. Once the water has completely boiled away (the drum is getting very hot now), carefully start pushing the sand around the drum at 6 inches (15 centimeters) high and 6 inches (15 centimeters) deep to restrict the airflow from the bottom (Figure 7-6).
11. Once flames stop coming out of the top, you can place the sheet metal you're using for the damper over the 6-inch (15-centimeter) hole. Also put some sand on top of the metal sheet to help hold it down and seal it even better (Figure 7-7).

Figure 7-3

Figure 7-4

Figure 7-5

Figure 7-6

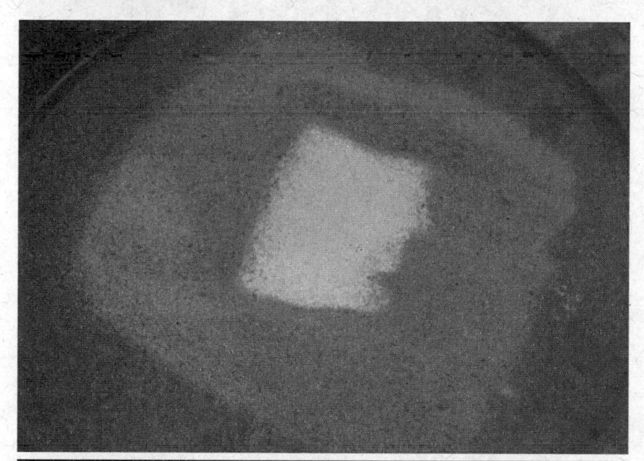
Figure 7-7

12. Let the drum cool for a few hours, and remove the sheet metal you used as a damper from the 6-inch (15-centimeter) hole. If there is a good bit of smoke still coming out, put the damper back on and wait longer. When there is little to no smoke, you're done.
13. Tip the drum over, and shovel the charcoal out. If it's not all charcoal, don't worry; the ashes add potassium and the unburned biomass can still compost (Figure 7-8).

Figure 7-8

How to Use Your Charcoal

Mix your charcoal with your compost at a maximum 10 to 1 (10 parts compost to 1 part charcoal). Regardless of how small the amount of charcoal, it still adds benefits.

Project 18
Make a Shopping Bag Composter

Using plastic shopping bags is one of the simplest ways to make compost. Composting in plastic bags takes very little attention and is a good alternative for beginners. Just be aware that bag composting favors anaerobic conditions because the amount of oxygen is limited, so the composting process is very slow (6 to 10 months). It is also slightly acidic, so you add the lime to counteract that. You'll need kitchen waste, water, and lime. The bags also need to be kept in a basement or warm garage to encourage decomposition during cold weather. No turning is required, and additional water won't need to be added after the bags have been closed. You can find the lime at local garden centers.

WHAT YOU'LL NEED

- Dry animal manure (composted is fine)
- Compostable materials (kitchen waste)
- Water
- Lime
- Plastic shopping bags (nonbiodegradable so as to not break down, fail, or spill)
- A dry location that doesn't get close to freezing temperatures
- A 2-cup (500-milliliter) measuring cup (Figure 7-9)

Figure 7-9

Let's Start

1. Fill the bags with kitchen waste until about two-thirds full (Figure 7-10).
2. Add lime to each bag, spreading it all around on the top layer (Figure 7-11).
3. Now add 1 cup (250 milliliters) of water to each bag and close it tightly (Figure 7-12).
4. After about 6 months, check the contents of the bags, and begin to use the compost if it resembles black dirt. If not, you can let it sit

Figure 7-10

Figure 7-11

Figure 7-12

Figure 7-13

longer or just speed up the composting by adding the material to your conventional aerobic compost pile (Figure 7-13).

TIP If odors become a problem, add more lime.

Project 19
Solar-Powered Bag Composter

This project is another type of anaerobic reduction (non-oxygen-loving microbes) of biomass called *fermentation*. We use black plastic bags to capture the sun's energy to help heat and speed up the composting process even further. Fermentation is a chemical change brought on by the action of microscopic organisms such as yeast, molds and bacteria. The souring of milk, the rising of bread dough, and the conversion of sugar to alcohol are all examples of fermentation. Alcohol is produced when yeast enzymes break up sugar into roughly equal parts of alcohol and carbon dioxide gas. Everyone may be familiar with baker's yeast, but most yeast occurs wild in nature and grows on plants and animals, by which it is dispersed through the air and water. Yeast spores are everywhere, and if they get a chance, they will gladly ferment your biomass waste, feeding off the waste's sugars, starches, and cellulose fibers. There are thousands of strains of wild yeast, and with the help of their enzymes, you can make a supercompost that is rich in helpful enzymes. Enzymes in the soil can degrade complex organic compounds into nutrients for plants and contribute to carbon turnover, providing energy for plant growth. Carbon-degrading enzymes are abundant in all healthy soil. By adding or increasing enzymes in soil, we are naturally helping in the cultivation of plants for food, energy, and industry.

Fermentation is an anaerobic conversion of sugars to carbon dioxide and alcohol that breaks down complex organic compounds into simple substances while producing enzymes. The alcohol produced during fermentation is called *ethanol*, like that found in wine, beer, and liquors. This ethanol is beneficial in helping to keep competing microorganisms such as black mold and other less beneficial microorganisms at bay. The fermentation process speeds up the production and reproduction of the enzyme proteins we want and also forms simple plant nutrients.

WHAT YOU'LL NEED

- Active dry yeast (the stuff sold for making bread)
- Compostable materials (kitchen waste, grass, leaves, and paper)
- Water
- Sugar
- Heavy (3-mil-thick) black plastic 40- to 50-gallon trash bags (nonbiodegradable so as not to break down, fail, or spill)
- A sunny location with warm days
- A 2-cup (500-milliliter) measuring cup
- A 5-gallon bucket
- A rubber band (Figures 7-14, 7-15, and 7-16)

Figure 7-14

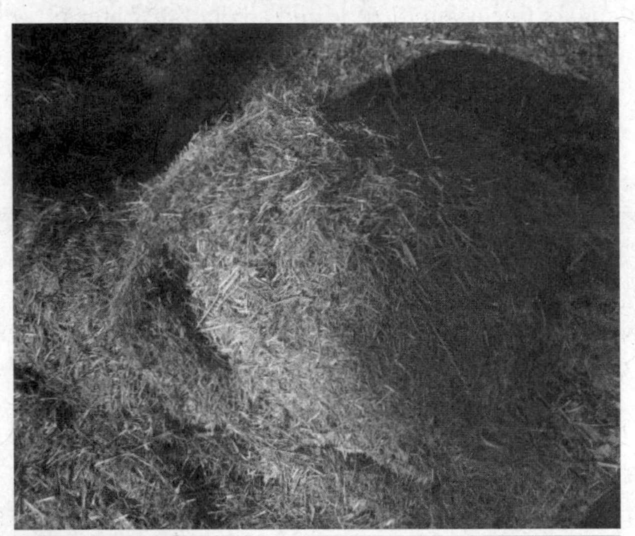
Figure 7-15

Let's Start

1. Find a nice sunny spot to work with your black plastic bags (black collects heat from the sun).
2. Add five 5-gallon (18-liter) buckets of biomass (leaves, grasses, and kitchen waste) to each bag. Be careful not to punch a hole in the bags. You don't want leaks. Fill each bag where you are going to let it sit in the sun so that you don't need to move it again (Figure 7-17).
3. Now fill the 5-gallon bucket about two-thirds full of room-temperature water, add 5 pounds (2.26 kilograms) of sugar, and stir it well to dissolve the sugar as best you can (Figure 7-18).
4. Add 5 tablespoons (150 milliliters) of dry active yeast to the water and sugar solution, and mix it all together (Figure 7-19).
5. Pour the sugar, water, and yeast solution into the bag.
6. You want the contents of the bag to be soaking wet, so if you used biomass with a lot of surface area, such as leaves, you may need to add more water.

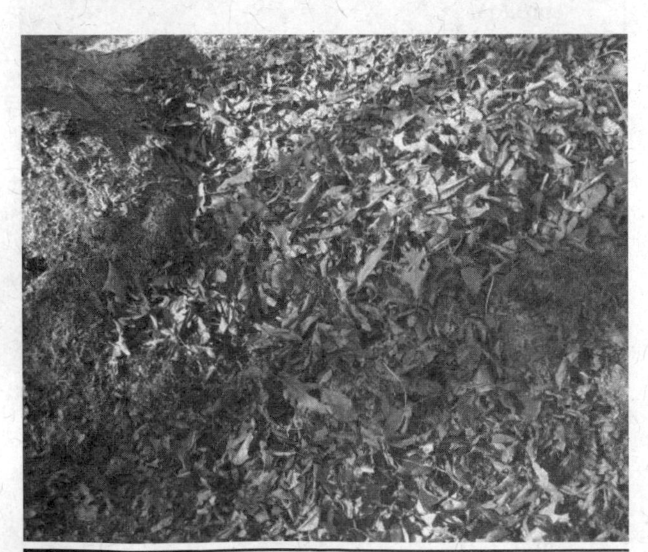
Figure 7-16

Figure 7-17

Figure 7-18

Figure 7-19

7. Once everything is soaking wet, grab the bag and loosely bundle it together in your hand.
8. Slowly squeeze out as much of the air as you can from the bag.
9. Now close the top of the bag using a rubber band to hold it shut, but not so tight that gases are prevented from escaping. During fermentation, CO_2 is created, and the bag will puff up as this happens. You don't want the bag to burst, so having a relatively loose seal is important. At the same time, you don't want outside air and other microbes getting into the

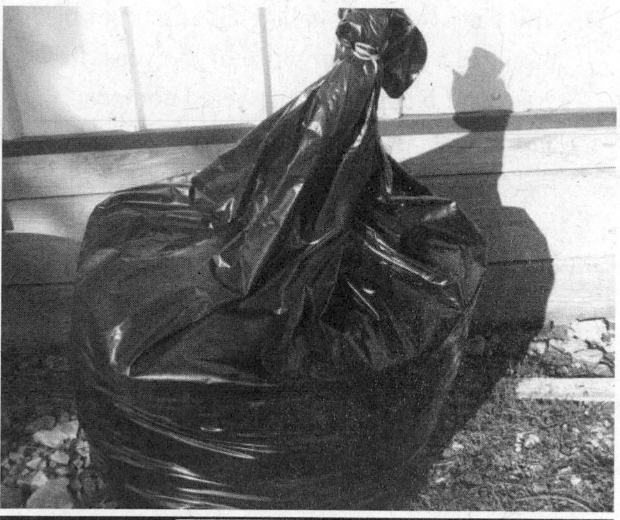
Figure 7-20

bag, so sealing it too loosely is also not a good thing (Figure 7-20).

10. Let the bag sit in the sun soaking up the sun's heat during the day and holding it at night, and in two to three weeks, the fermentation should be finished. If not, add more sugar, and give it more time.

How to Use It

1. We first need to separate the liquids and the solids, so by hand (wear gloves if you want), pull out as much of the solids as you can and put them in the bucket. When the bucket is full, put its contents in with your other compost or soil, and blend it in well. Repeat this step until the bulk of the remaining solids is removed.
2. Pour the liquid contents into your empty bucket, and use them as a superplant food by mixing them first with water at 1 tablespoon (30 milliliters) of compost liquid enzyme per 1 gallon (3.8 liters) of water. Spray it and pour it on your plants to make them healthy and to discourage pests.

3. You can also use 1 gallon (3.8 liters) of the liquid as a culture in place of the yeast for your next bag of solar-powered compost.
4. The liquid also can be used to boost water-based cleaners around your home, just like the cleaning enzymes in Chapter 8.

Project 20
Build a Barrel Composter

Making compost in a barrel is like shooting fish in a barrel—you can't miss. This is a great way for people with limited space to make compost. It's an aerobic (oxygen-loving) microbe method of composting, so it is quite quick at doing the job.

WHAT YOU'LL NEED

- Compostable materials
- Manure
- Water
- A clean 55-gallon plastic drum with a removable lid
- A ½-inch drill bit
- A drill

Let's Start

1. Find a plastic barrel that's at least 55 gallons in size with a secure and easily removed lid. Be sure that the barrel wasn't used to store toxic chemicals.
2. Drill nine rows of ½-inch holes over the length of the barrel to allow for air circulation and drainage of excess moisture.
3. Place the barrel upright, fill the barrel three-fourths full with organic waste material, and add about 6 inches of manure as a nitrogen source.
4. Add some water to make the compost moist but not soggy. Close the lid, and place the barrel on its side. Give it a good rolling to mix the contents well.
5. Every few days, roll the drum around your yard to mix and aerate the compost.
6. The compost should be ready in two to four weeks if things work well.

Diagnosing Composting Problems

1. If the pile is producing a bad odor, it may be too wet or packed too tightly or both. Turn the compost to loosen it and allow better air exchange in the pile. If the compost is too wet, also turn the pile, but at the same time, add new dry materials.
2. If no decomposition seems to be taking place, the pile is too dry. Moisten the materials, and turn the pile.
3. If the compost is moist, sweet smelling, and has some decomposition but seems slow, try mixing in another nitrogen source such as fresh grass clippings, manure, or a natural fertilizer.

Project 21
Build a Raised-Bed Composter That Turns into a Garden

A raised-bed garden can be built just about anywhere there is sunlight, and if you use compost to make up most of its growing medium, it is even better. A raised-bed garden saves time and energy; it also naturally suppresses weeds and creates a viable growing base for plants. You'll need a frame to hold the raised bed above the existing ground. Weeds and grass will be unable to penetrate beyond the frame because it makes an edge that's hard to breach. A good-sized raised-bed garden is 8 feet (240 centimeters) long by 4 feet (120 centimeters) wide and 16 feet (40 centimeters) high. You should limit the need to walk on or in your raised-bed garden, so a bed this size makes it easy to reach the middle. Start this project several months ahead of when you want to use it as a garden because it can take several weeks for the compost to be ready. Two beds of this size could supply a whole family with fresh vegetables.

WHAT YOU'LL NEED

- Natural fertilizer (high nitrogen)
- Newspapers or cardboard
- Compostable materials
- Water
- Manure
- Lime
- Soil
- Concrete papier-mâché landscape blocks
- Concrete papier-mâché slump
- Hard-toothed rake
- Hammer
- Shovel

Let's Start

1. For the frame we'll use our concrete papier-mâché blocks because we can and they're affordable. Build the frame in the same way you would've built the walls for Project 10 in Chapter 6.
2. Inside the frame, lay newspapers or cardboard about ½ inch (1 centimeter) thick. Make sure that the layers overlap well to positively smother out the weeds or grass underneath. The paper or cardboard will rot and eventually become part of the compost layer. Use wet paper if the wind is blowing hard so that you don't have chase it down when it blows away.
3. Add a layer of weed-free grass clippings or leaves on top of the paper/cardboard layer. Sprinkle or spray some organic fertilizer or add a layer of manure to help with the breakdown.
4. Add another layer of weed-free grass clippings or leaves, and again spread a little fertilizer or add a layer of manure.
5. Give the raised bed a little water from time to time to encourage the decomposition of your organic biomass.
6. Keep adding these layers until the pile is a little higher than your frame walls. In a few weeks, the level will drop, and the level will need to be maintained by an infusion of soil or finished compost.
7. Once every few days, use your hard-toothed rake to turn the length of the pile. If you want to speed things up, add a little lime from time to time to balance the pH.
8. In as little as a few months your garden can be ready for planting if everything goes well.
9. Maintaining your bed is easy because the weeds from the original soil will have been smothered by the paper and cardboard, and the weeds that do find their way into your raised bed are going to be shallow-rooted and easy to pull out. The height of the bed gives you a good set for working the garden, saving your back.

Project 22
Build a Solar-Powered Compost Tea Brewer

Compost tea is the perfect nutrient boost for flowers, herbs, house plants, and vegetable gardens. If you are not currently making and using your own compost tea, then you're probably unnecessarily spending extra money for liquid fertilizer. The nutrients and minerals for compost tea are readily absorbed by plants, making for greener leaves, bigger, brighter blooms, and more desirable fruits and vegetables. To make compost tea, you basically soak compost in water for a few days and supply the water with oxygen. The brewing compost will give off nutrients and minerals to the water, making your tea, which is then drained off for use in feeding plants. To find a solar-powered aeration pump to use in your brewer, try local garden or pet stores that supply equipment for koi fish ponds, or search online for "solar-powered aeration pumps," and choose the one you want. If you can't find a solar-powered pump, you can use a conventional air pump from a fish aquarium.

WHAT YOU'LL NEED

- Rain or unchlorinated water
- Sulfur-free molasses
- Mature compost
- Air line for the solar-powered air pump
- A recycled hole-free cotton sock
- A 2-cup (500-milliliter) measuring cup
- A 5-gallon bucket
- A mixing stick
- A funnel

Let's Start

1. Find a nice sunny spot, and place the air line in the 5-gallon bucket. Make sure that it reaches the bottom of the bucket.
2. Add 4 cups of your mature compost to the same 5-gallon bucket. Make sure that the end of the air line is covered.
3. Add water, filling the bucket to within a few inches of the top.

TIP **If you are using tap water, it may contain organism-killing chlorine, so let it sit overnight before you add it to the compost.**

4. Add 1 tablespoon (30 milliliters) of molasses to provide a food source for the beneficial microorganisms you want to reproduce.
5. Turn on the solar-powered air pump, and let the contents of the brew bubble for two to three days. Stir the brew occasionally with your mixing stick to mix the compost and release the microorganisms from hanging onto the compost's solid particles.
6. After brewing the mixture, you need to strain it. Use the recycled cotton sock and your funnel, and strain the tea into another bucket.
7. Put the compost solids back into the compost pile or the garden.

TIP **If the compost tea smells bad, dump the mixture back into your compost pile. It didn't get enough oxygen and has spoiled. The compost tea should smell sweet and a little earthy.**

Compost Tea Tips

- When making a compost tea, always use un- or dechlorinated water. You don't want to kill off the good microbes.
- Use compost to make a tea that has a pleasant smell.
- To ensure the greatest variety of microbes, use molasses, sugar, carbohydrates, and proteins to feed and strengthen the aerobic organisms in your tea.
- If foam is produced that is several inches thick, just add a few drops of vegetable oil to break the surface tension of the bubbles and stop the foam formation.
- If you are not going to use the liquid compost tea right away, continue to brew it with the aeration to keep it from spoiling and becoming anaerobic. Even aerated teas last only 2 days.
- In between batches, clean your equipment to stop the buildup of biofilms such as *Escherichia coli* bacteria.
- Sprinkle or spray your compost tea onto foliage and soil around each plant every two weeks.

CHAPTER 8

Clean with Green Projects

Today's household cleaning products may contain potentially hazardous and nonrenewable ingredients such as organic solvents and petroleum-based chemicals. As a result, many people are rediscovering the safer and just-as-good natural household cleaning solutions that our parents and grandparents used. Old-fashioned homemade cleansers are simple to put together and can be made in large quantities. You might want to make and label large batches of these cleansers because they are so inexpensive and easy to store.

Cleaners 101

Cleaners come in many types—abrasives, oxidizers (bleaches), acids, salts, surfactants, and mechanical ("elbow grease"). No matter what you want to clean, more often than not you will employ a solvent of some kind to help make cleaning happen. Cleaners use solvents that can dissolve, suspend, or extract other materials, usually without changing either the solvent's or the other material's own chemical bonds. Solvents are called *organic* or *inorganic* depending on whether the solvent contains carbon as part of its makeup. For example, alcohol is an organic solvent because it contains carbon, whereas water is an inorganic solvent because it is carbon-free. Thus the words *organic* and *inorganic*, when used as chemical terms, have nothing to do with a solvent being from a green source. In general, earth-friendly or not, hydrocarbon and oxygenated solvents are "carbon-containing" organic solvents that can dissolve many materials effectively and are often mixed with the inorganic "non-carbon-containing" solvent water. To understand how solvents help to clean, think about how humankind washed clothes about a hundred years ago with basically just soap, water, and a bucket containing a washboard. The clothes were cleaned by scrubbing them against the washboard, and the mechanical force of rubbing against the washboard would help the solution of soap and water to remove the dirt. Today's washing machines provide their own mechanical force, and the solvents we use (detergent and water) help to reduce the force required to remove or soften soils and dissolve unwanted materials. Solvents help to save energy in a roundabout way.

A surface cleaner without solvents can cause streaking because soil particles actually may settle back to the surface before they can be wiped away. The types of cleaners that rely on solvents include surface cleaners, window cleaners, and floor polishes.

Solvents also help with disinfectant "germ killing" and stain-lifting tasks by lowering surface tension on the material being cleaned. This feature, working with other ingredients in the product, assists disinfectant or water to get into the nooks and crannies of what is being cleaned to lift, remove, or saturate unwanted germs or stains.

Solvents are found in most home cleaning products to improve cleaning efficiency and make homes both easier to clean and more hygienic. Therefore, when you clean, you're diluting, bleaching, scrubbing, neutralizing, and making something "foreign" adhere to a cleaning cloth or suspend in rinse water to be removed. With less time spent on cleaning and scrubbing, you now have more leisure time and use less energy.

Anatomy of a Cleaner

Green, natural, earth-friendly, and even conventional cleaners on the shelves of stores contain or use water as a main ingredient that allows them to perform. When you look at the ingredients listed on the label of a liquid-based cleaner, you see that water is the first ingredient listed. Manufacturers don't print the exact percentage of the ingredients used in their products to make it easy for us, but it is common practice at least to list the most abundant ingredients first. Thus, if water is listed first, you can bet that it is over 90 percent of the cleaner's contents. This is not a bad thing; water is an inexpensive and powerful solvent. The water isn't diluting the ability of the cleaner; rather, it's an important part of that cleaner. However, why would you pay for mostly water? Plus, how about all the energy that went into shipping the added weight of the water in that product? The amount of energy spent shipping that cleaner made up mostly of water per household may be small, but times by millions of homes and worldwide, the process needlessly spends millions of dollars on something we already have in our homes. The added water and its packaging are a convenience, making the cleaner ready for us to use as a spray or a measured liquid. Moreover, the bulk helps to make the product look better and thus more marketable. Let's face it, I would be hard pressed to pay several dollars for just 650 milliliters (2 ounces) of a cleaner to which I had to add my own water. Even powder cleaners are diluted with less inexpensive fillers to give them mass and marketability. To be fair, the fillers used in cleaning products still help them do their jobs to an extent. The remaining 10 percent or less of a cleaner's ingredients, whether the cleaner is for your toilet, laundry, carpet, or even windows, is formulated from varying amounts and combinations of surfactants, organic (carbon-based) solvents, acids, salts (alkalis), oxidizers (bleach), and abrasives (minerals).

Laundered Money

It's easy, therefore, to understand why so much money is spent promoting cleaning products—they are very profitable for manufacturers and retailers. We as consumers spend hundreds of dollars each year on cleaning products. However, just knowing that 90 percent of what we bought is something that we already have in hand and is inexpensive or nearly free, it's no wonder that we can feel as if money is being washed right out of our pockets.

Cut Your Cleaning Cost by 90 Percent or More

Knowing that 90 percent of what we buy consists of inexpensive or free ingredients, it's easy to see that saving money making our own cleaners is a viable option. If you also can use safer ingredients that are more friendly to people, your pets, and the earth you live on, the benefit is even greater.

Home Basic or Base 50:1 Concentrate

What we are going to make is a three-part concentrated "base" or "basic" clean-all cleaner that can be activated for different cleaning tasks by adding other ingredients. We're essentially making a concentrated "basic" cleaning solution that provides us with a "base" to accomplish a specific cleaning task by means of other additives that we mix with that base. Let's begin by putting together a plan for collecting or making our basic ingredients toolbox. Our goal should be eco- and user-friendly as well as consisting of common and somewhat easy-to-find things that we have in, near, or around our homes. The most important ingredient that we need to make our base is mineral-free, or soft, water. This is not just plain tap water out of our faucet (unless you have a water softener or reverse osmosis unit); this water will helps us to prepare a proper and predictable concentrated base cleaner. The next ingredient we need for our base is a surfactant (soap); this ingredient makes our solvent(s) better at making contact with the item we wish to clean, breaking the chemical bonds between the dirt and the object, and finally helping to lift away the dirt, grease, and grime. Last but not least, we need a stabilizer to preserve and help to keep our base in a uniform mixture so we can get consistent and predictable outcomes as we mix other ingredients to use our base cleaner for many different applications.

NOTE **I use the word *base* to describe our *basic* concentrated cleaner, and in this context, it refers to the cleaner itself. However the word *base* from a scientific standpoint is often used to identify something as an alkaline, salt, or opposite of acidic.**

NOTE **The base concentrate can make almost 150 gallons of diluted cleaner. From a savings standpoint, this is about $3,000 worth of cleaner when compared with equivalent store-bought cleaners.**

Basic Ingredient 1: Wetter Water

You've seen me use the term *soft* or *mineral-free water*. To understand what "soft" water is, we need to examine "hard" water. Hard water is any water containing dissolved minerals. Soft water, or mineral-free water, is as common as rain. It is made by running tap water through a membrane filter, distilling it, or submitting it to an ion-exchange resin, and the result is a treated water in which the only remaining mineral is positively charged sodium ions. By passing hard water over an ion-exchange resin, the untreated water flows over the resin surface, and the dissolved minerals (ions) in the water stick to the resin, causing the exchange (release) of the dissolved sodium (ion) from the resin's surface. Sodium goes into the water, whereas the other minerals stay with the resin. One of the reasons people soften hard water is that hard water shortens the life of plumbing pipes and lessens the effectiveness of cleaning agents. When hard water is heated, the minerals precipitate out of solution, forming scale in tea kettles or in pipes. In addition to slowing the flow and clogging the pipes, mineral scale found in water heaters prevents efficient heat transfer, so a water heater with scale will use a lot more energy to produce hot water. Cleaners are also less effective with hard water because the water reacts to form calcium or magnesium salts with the organic acid of the soap (scum). These salts are insoluble and form a grayish soap scum. We don't want our homemade cleaners to cause soap scum or leave mineral streak marks, so we must use mineral-free, or soft, water in our cleaning solutions. To be fair, some minerals in water aren't all bad. Certain natural mineral waters are highly sought after and even bottled for their flavor and claims of health benefits.

Basic Ingredient 2: Surfactant

Nearly all compounds fall into one of two categories: *hydrophilic* ("water loving"—anything that will mix with water) and *hydrophobic* ("water hating"—oil and anything that won't mix with water). When water and oil are mixed, they separate. Hydrophilic and hydrophobic compounds just don't mix.

When we use a soap-and-water solution, the soap molecules work as a peacekeeper between the hydrophilic and hydrophobic compounds, making the polar water molecules and nonpolar gunk, grime, or glob molecules easier to remove. Since soap molecules have both properties of nonpolar and polar molecules, the soap can act as an *emulsifier*. An emulsifier is capable of dispersing one liquid into another immiscible liquid. This means that while oil (which attracts dirt) doesn't normally mix with water, soap can suspend the oil/dirt in a way that allows it to be rinsed away (removed).

Basic Ingredient 3: Stabilizer

A *stabilizer* can be a natural or synthetic chemical that is added to foods, pharmaceuticals, paints, biological materials, and cleaners to prevent their premature decomposition by microbial growth or by undesirable chemical changes. In our cleaner, we are going to use boric acid or borates, which are the naturally occurring salt compound of the element boron. Borates are widespread and abundant in soil, water, and food. These salts are often recommended as the least-toxic pesticides for killing insects, mites, algae, fungi, and molds. Another advantage of using borates as a stabilizer is that they also have a surfactant effect and thus aid in keeping our ingredients from separating when not in use.

Base "Basic" Ingredient Recap

1. Mineral-free, or soft, water (solvent)
2. Surfactant (soap)
3. Stabilizer (preservative)

How to Collect Soft, or Mineral-Free, Water

If you don't have a water softener distillation unit or a reverse-osmosis ion-exchange filter to make your soft water, it can be bought at a number of stores. Read the labels on bottled water, looking for such words as *reverse osmosis* or *distilled*. Beware of spring-fed or treated waters, however, because they still contain minerals. Another source of mineral-free water that's nearly free is rainwater or melted snow. Snow is easy; just pick it up, put it in a clean waterproof container, and let it melt. Rain can be caught in a clean waterproof container by just setting it outside in a steady rainfall. For those of you who want more rainwater, you can build a catch that sits under your roof's downspouts to capture, filter, and store rainfall.

You can buy attractive wooden rain barrels with an old-fashioned appearance or plastic rain barrels that might not be as pretty but still get the job done. Manufactured rain barrels come with faucet fittings to allow you to hook up a garden hose so that you can direct the water wherever you choose. It's as simple as placing it under a downspout. Your rain barrel will fill up quickly in a heavy rain.

Project 23
Make Your Own Rain Barrel

WHAT YOU'LL NEED

- A clean plastic 55-gallon barrel, preferably one that was used in a "food grade" situation
- A saber saw
- A couple large bricks, blocks, or rocks
- A drill
- A 1-inch (25-millimeter) drill/spade bit
- A garden hose
- A shutoff valve for the hose
- A shovel
- Eye protection (Figure 8-1)

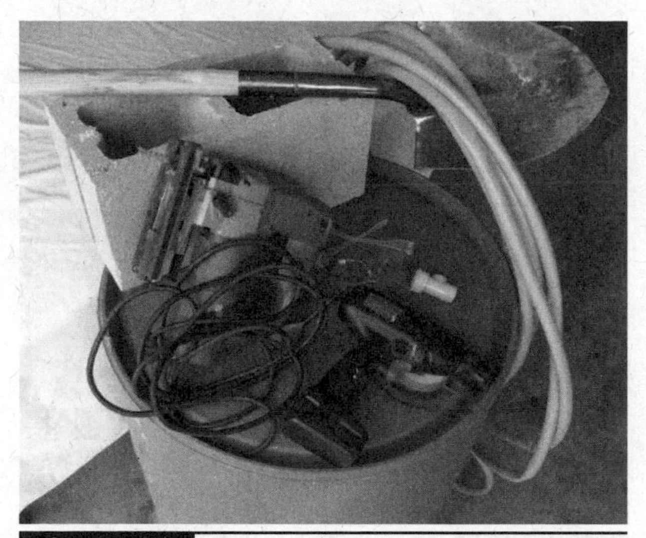
Figure 8-1

Let's Start

1. Find a clean plastic 55-gallon barrel. Wash and rinse it a few times with soap and water to make sure that it is clean.
2. Locate the downspout you want to use to capture rainwater. Make sure that the spot is level. Dig out a level base with your shovel if need be.
3. Place the bricks, blocks, or rocks on the leveled spot. You also want the blocks to be level because this is where the barrel is going to sit. You want the barrel to sit high enough above ground that you can place a water bucket under it [about 1 foot (30 centimeters)] (Figure 8-2).
4. If your barrel does not have an opening at the top, use a saber saw to cut one to the size of your downspout as close to the edge of the barrel as you can (Figure 8-3).
5. Set the barrel on top of the leveled blocks, and fit the downspout into the hole you just cut in the barrel's top (Figure 8-4).

Figure 8-2

Figure 8-3

Figure 8-4

6. Now drill four or more holes on the side of the barrel near the top and facing away from the house. These holes let the water spill out slowly if the barrel gets too full (Figure 8-5).
7. Drill a 1-inch (25-millimeter) hole opposite the side of the downspout. Fit the female end of the hose into the hole you just drilled. Put nearly the full length of the garden hose into the barrel. Tie a loose loop knot at the end of the hose so that it doesn't also fall into the barrel through the hole (Figures 8-6, 8-7, and 8-8).

Figure 8-7

Figure 8-5

Figure 8-8

Figure 8-6

How to Use Your Barrel

When your barrel is nearly full of rainwater, the hose that is tucked inside is also nearly full. We now want to keep air from getting into the hose so that we can produce a vacuum and use the hose to syphon water out of the barrel. Put the shutoff valve on the hose. Open the valve, and put as much of the hose into the barrel as possible. Then close the valve to stop air from pulling back into the hose as you move the hose out of the barrel. Pull the hose out of the barrel slowly, and lower the end that you are holding below the bottom of the barrel. If you kept air out of the hose, the syphon action should be good to go. If not, repeat the preceding steps, and check to make sure that the hose is not pinched off in the barrel. If you need rainwater before the barrel is full, have someone help you tip the barrel, and pour some water out. You also can drill a small hole in the bottom of the barrel, and plug it with a large wood screw.

How to Make or Find Your Surfactant (Soap)

Lye soap was once made from the fatty acids found in animal fat or even vegetable fat that was rendered and used just for soap making. Today's soaps are made from petroleum because they can be superconcentrated and cost very little to make. Regardless of whether such soaps are earth- or user-friendly, most consumers want cheap and powerful soaps.

Lye to Me

Lye is an extremely caustic agent, so soap makers had to be careful to have just the right concentration. Too much lye causes the soap to burn the skin, and too little keeps the soap from hardening. An egg was used to determine the proper amount of lye. When an egg was floated in the mixture, the amount of lye was correct when only the tip of the egg was seen above the surface. To make lye soap, lard and lye were mixed together over an open fire and stirred for hours until the mixture thickened. People once used lye soap to clean everything from their faces to their laundry. Today, many people still like to buy and use lye soap. For our cleaner base, we want a natural soap that is free of additives such as oils and fragrances. Natural soaps are still inexpensive, so keep this in mind before you go to the trouble of making your own.

To Buy or Not to Buy

I found natural soap noodles for around $3 per pound. This type of soap is used as a neutral starter for people who make their own soap or crafts. Soap noodles are shredded soaps, and they are the way to go if you want to make your own natural soap but do not want to use lye (sodium hydroxide). You want soap noodles that are pure soap with nothing else added; they are 100 percent soapified natural oils.

Project 24
Making Soap (Surfactant)

Soap making is thousands of years old, and one of the best and most often used soaps is called *lye soap*. Even today's soaps are a type of lye soap, but today they are made from mass-produced sodium or potassium hydroxide instead of coming from wood ashes or other caustic mineral such as sodium carbonate and lime. Caustic lye used to make soap often was made from the ashes from wood stoves. For hundreds of years, ashes for making soap were kept in wooden bins. When it came time to make lye soap, people poured water through the ashes and siphoned off the liquid lye.

Where to Find Lye (Sodium Hydroxide)

Lye is an everyday drain cleaner, but for soap making, it must be pure. In many states, however, you can no longer get simple drain cleaner because of its use in making the illegal drug methamphetamine. Thus it's best to do an Internet search to find lye and get it online. Just type in "soap making supplies" or "lye for soap making," and select the vendor you want. Lye can be dangerous, yes, but life is fraught with danger. Stoves are dangerous. Crossing the street is dangerous. Get over it. If you follow the rules and proper handling procedures, you will be fine.

The soap-making process creates a chemical reaction called *saponification*, after which fat is no longer fat and lye is no longer lye—together, they have become soap. If it doesn't have lye in it, it's not soap. Some producers of "homemade" soap, understanding people's aversion to lye, list sodium hydroxide as an ingredient instead of lye. Sodium hydroxide *is* lye. Many commercial "soaps" do not list lye or sodium hydroxide as an ingredient. However, if the cleansing bar you buy does not contain lye, it is *not* soap—it's detergent.

Where to Find Fats for Making Soap

All grocery stores stock fryer oils and lard. Basically, the best fat/oil to use is the one that's the most unhealthy for your diet. Hydrogenated oils work great (restaurant fryer oils are partially hydrogenated), lard is wonderful, and even waste fryer oils left over from cooking work well.

Used Restaurant Oil (Fat)

Go to a local restaurant that isn't part of a larger chain, and ask the manager if you could have some of the waste oil when the restaurant changes out its frying oil. Let the manager know that it's okay to put the oil back into one of the empty jugs the new oil comes in. Tell the manager what you're doing with it, and most managers will be more than happy to help. If you don't have any luck getting it for free, offer the manager a few dollars; this is still very affordable when compared with the cost of new oil. Do not take oil from outside grease disposal and recycling containers. Once the oil is in that container, it's someone else's property, and taking it is stealing. Even if the restaurant is paying to have it hauled away, it's still stealing once the oil is brought out the back door. To use it for making soap, you want the oil to be clear and lightly colored (like iced tea). Leave the container of oil in a warm place, and let any crunchies or solids settle to the bottom. If you need to filter it, pour the oil into an old hole-free cotton sock, or put a terry cloth towel over a bucket and pour the oil through the towel, catching the oil in the bucket.

TIPS

- **Wear a long-sleeve shirt, full shoes, and trousers. Wear chemical-proof gloves, an apron, and eye protection such as goggles or a full-face shield. Always have running water available to wash away spills.**
- **Never use aluminum in the soap-making process.**
- **You must work in an area away from children and pets where it is safe to make the soap.**
- **Do not touch lye. Use a scoop, wear gloves, and protect your eyes.**
- **Do not leave the lye solution unattended.**
- **Keep children and curious pets away.**
- **Open a door or window, go outside, or activate an exhaust fan to avoid inhaling dangerous fumes.**
- **If you get lye on your skin, rinse it immediately.**
- **Vinegar counteracts lye because it is an acid.**
- **If you spill lye solution on the countertop, wash it immediately with vinegar. It is best to cover countertops with plastic for safety and to avoid a big mess.**

CAUTION **Always add the lye to the water. (Otherwise, you may create a dangerous exothermic reaction.)**

CAUTION **This is not a soap with which you want to wash your body or face. It is meant for use as a household cleaner.**

WHAT YOU'LL NEED

- An accurate scale (it needs to weigh down to 1/10 of an ounce)
- A stainless steel or enameled pot
- A plastic or wooden spoon with holes or slots
- Two plastic pitchers; at least one needs a lid
- Rubber gloves

WHAT YOU'LL NEED

- Eye protection
- A hot plate or stove
- A heavy apron
- Measuring cups
- A thermometer that reads as low as 90°F and higher than 200°F
- A stick blender (used for making milk shakes in a glass)
- A corrugated box of approximately 8 × 8 × 9 inches (a shoe box)
- A small size plastic trash can liner or plastic shopping bag (Figure 8-9)

Figure 8-9

Alternatives to Waste Fryer Oil

- 1 pound of oil (any type will do, the cheaper the better)
- 3 pounds of lard
- 1 pound of coconut oil (semisolid at room temperature)

CAUTION Never use aluminum with caustics such as lye. It can cause undesirable effects and also be dangerous. Use only stainless steel, enamel-ware, glass, Pyrex, wood, or plastic in the soap-making process.

TIP Some pots and pans may look stainless but are in fact not, so if a magnet won't stick to it, be aware that it might be aluminum. Some grades of stainless also are not magnetic, but I'm unaware of any of those types being used for pots and pans.

Let's Start

TIP Making soap base is a cumbersome task. This method is going to appeal only to the chemistry-minded type of person. If you can't make brownies from a box mix turn out right, you'd better order some shredded soap rather than attempting to make it yourself.

TIP Put on your gloves and eye protection. Read the warning label on the lye. It is a caustic and dangerous substance. It makes wonderful soap, but it is not your friend and can hurt you.

1. Weigh an empty lightweight plastic container first, and adjust the scale to zero. Weigh out 32 ounces (907 grams) of ice-cold soft or mineral-free water (Figure 8-10). Never use hot water to mix with lye; it will erupt like a volcano.
2. Weigh another empty lightweight plastic container first, and adjust the scale to zero. Weigh out 11.2 ounces (318 grams) of lye, and then pour the lye carefully and slowly into the ice-cold water. Ensure adequate ventilation; outdoors is best. Lye pulls moisture out of the air, so it has a lot of "static cling." Spills are easy to do. You'll notice that the lye will react with the cold water, and it

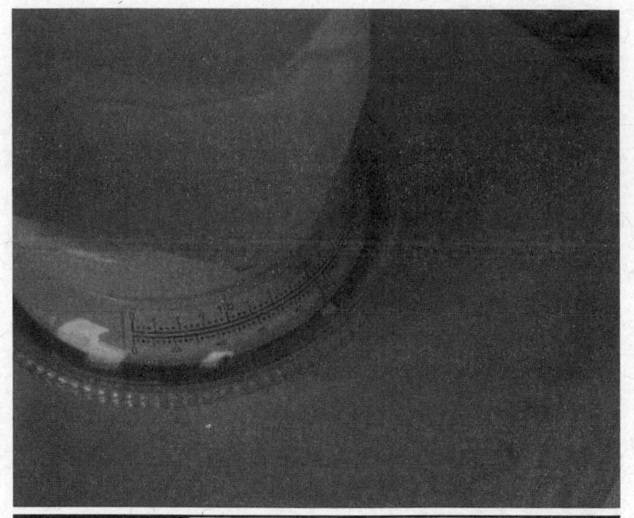

Figure 8-10

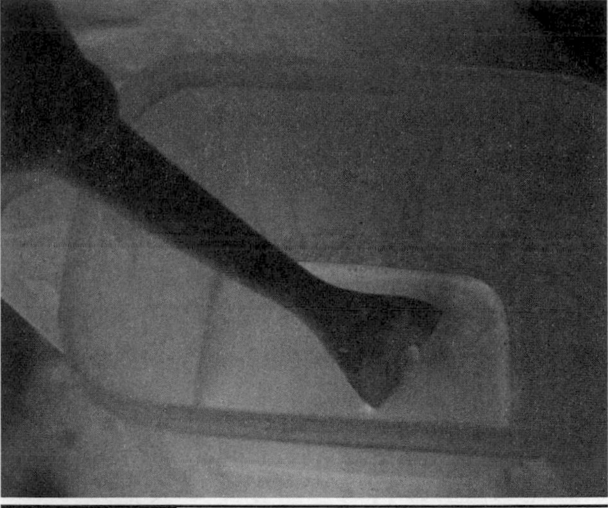

Figure 8-12

gets very hot. It'll also give off a gas; this is why you should be outside. Don't breathe the fumes. When the mixture is stirred and the crystals have dissolved, cover the lye solution and bring it back inside. Let your lye sit in a safe place (out of the reach of people and pets) until it cools to room temperature. This will take 2 to 3 hours (Figures 8-11 and 8-12).

TIP You can make this lye solution the day before to give it plenty of time to cool.

Figure 8-11

CAUTION Lye solution is a very dangerous substance. It can cause death if ingested. Make sure that everyone in the house knows what it is and that it is not to be touched.

3. When the lye and water mixture has returned to room temperature, you're ready to start making your soap. Start by weighing out your fat.
4. Pour your weighed oil into a stainless steel or enameled pot. Put that pot on the stove to heat while stirring often. Keep a close eye on it because oil can reach temperature unexpectedly and somewhat quickly. Stir well before measuring the temperature. You want the temperature between 120 and 130°F (60°C) (Figure 8-13).
5. Make sure that the lye solution is near room temperature. Make sure to wear your gloves and goggles. Now carefully and slowly pour the lye solution into the oil. Hold your head back while pouring to avoid splashes.
6. Use a large wooden or plastic spoon with holes or slots, and carefully stir the mixture. Keep stirring for a few minutes. Then switch to the electric stick blender, or you can continue stirring with the spoon for about an hour if you like (Figure 8-14).

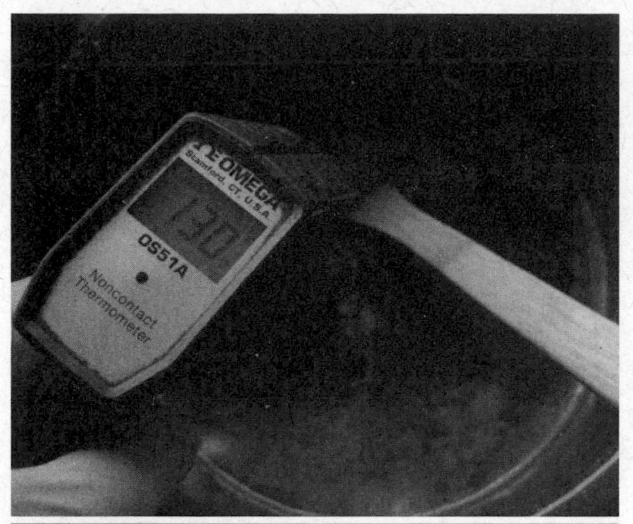

Figure 8-13

Figure 8-15

Figure 8-14

Figure 8-16

7. If you use a stick blender, pulse it on and off to help keep the mix from splattering all over the place. Blend with the stick for a minute, and then stir with a spoon for a minute, if you need to. The result should be a very thick liquid soap in 5 to 10 minutes or so. It will become thicker and more murky, cloudy, or muddy looking as time goes on if you weighed accurately and had your temperatures accurate (Figures 8-15 and 8-16).

TIP **The bowl and mixture will stay warm to the touch and may get even warmer owing to the sopifacation process. So don't be afraid of a little heat.**

8. Immediately pour the soap into the mold (the 8- × 8- × 9-inch box lined with a plastic bag). Since the sopifacation reaction is still taking place, even as it sets in the mold, it generates heat. If you put your hand on the side of the box, you will feel its warmth. Let it cool for 24 hours (Figures 8-17 and 8-18).

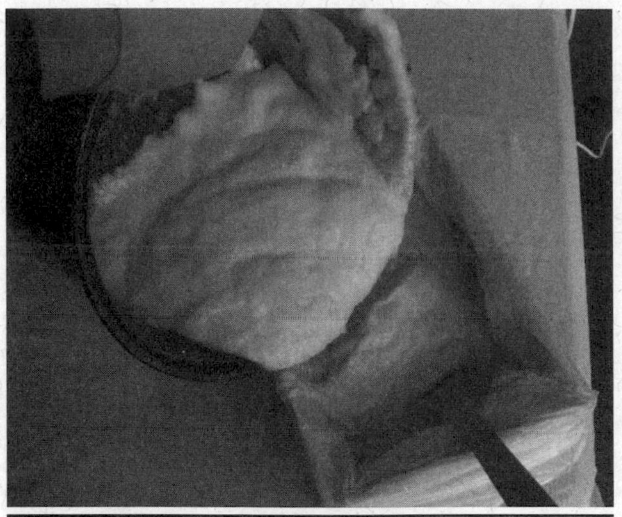

Figure 8-17

Figure 8-18

9. After 24 hours, the soap should be solid or semisolid, and it's good to go. However, if it is mushy (because of the quality of the oil), you'll need to move it back into the pot and add a couple tablespoons of plain old table salt. Put it on the stove, and reheat it slowly while stirring constantly. When it reaches 120 to 130°F again, remove it from the heat, and pour it back into the mold. Give it another 24 hours, and if it's still not solid or is semisolid, your weights of oil and lye may have been off. The best place for this failed batch may be your compost pile—a few ounces at a time over the next few weeks.

Project 25
Making a Concentrated Base "Basic" Cleaner

This is the basic or base cleaner concentrate that will be used to make many other household cleaners. The recipe (formulation) here makes around 3 gallons of cleaner concentrate.

WHAT YOU'LL NEED

- A stainless steel or enameled pot
- A large plastic bucket with a removable lid
- A plastic or wooden spoon with holes or slots
- Rubber gloves
- Eye protection
- A heavy apron
- A measuring cup
- A hot plate or stove
- 2 gallons (7.6 liters) of soft, or mineral-free, water
- The block of soap you just made in Project 24 or 5 pounds (2.2 kilograms) of natural fragrance– or oil-free soap or soap noodles
- A hammer
- A chisel
- A grater
- 25 fluid ounces (750 milliliters) of consumable 190 proof grain alcohol (stabilizer)
- 16 ounces (453 grams) of borax (stabilizer) (Figure 8-19)

Figure 8-19

Let's Start

1. Place the opened 4- to 5-gallon bucket in your sink or on a surface where you don't mind cleaning spilled cleaner.
2. Spread out some newspaper, and break your solid to semisolid soap into smaller pieces. Use your hammer and chisel if necessary. You can take pieces that fit in your hand and, while wearing gloves, grate them on a grater like you would cheese. If you bought the natural soap noodles, you can skip this step.
3. Fill the stainless steel pot or enameled pot with a couple quarts of the mineral-free, or soft, water [from the 2 gallons (7.6 liters) stipulated for this project], and bring it to a boil. Once it is boiling, add the soap pieces you just made or the soap noodles you bought, and stir. Reduce the heat, and add just enough water to about fill the pot (leave some space at the top of the pot so that the solution doesn't spill out as you stir). Continue to simmer the solution while you stir. You want to dissolve the soap completely in the water. When a fair portion of the soap has dissolved, pour as much of the liquid out into the large bucket as you can without any solid soap pieces falling out of the pot. Place the pot back on the heat, and add more of the remaining 2 gallons of water until the pot is nearly full again. Simmer and stir again until all the soap is liquified into the water, and add it to the bucket.
4. With the liquified soap in the bucket, you now need to add and stir in the rest of our water, borax, and entire 25 fluid ounces (750 milliliters) of 190 proof alcohol.
5. Once this is stirred and mixed well with all the borax dissolved, you can place the lid on the bucket and use the contents whenever you want.

How to Use It

The primary use for this base concentrate is to give us the building blocks for all the types of cleaners we use in our homes. For a general-purpose surface and floor cleaner, just mix 1 part of the concentrate with 50 parts water.

Project 26
Testing Your Base Cleaner's pH Level

Knowing where the pH falls for your base cleaner is important for making a pH-balanced mild laundry detergent or stain remover that's colorfast and safe for fabrics. The pH is basically a measurement of the hydrogen ion concentration (parts hydrogen). A sample of absolutely pure water has a pH of 7.0. The pH scale ranges from 0 acidic (acids) to 14 base (alkaline). A pH of 7.0 is considered a neutral pH. Your mild detergent or stain remover should be near neutral on the pH scale, between 6 and 10.

NOTE The pH of your base cleaner can vary owing to the type and quality of the oil you used, as well as how precise your measurements were when you made it. Most likely it will be a little on the (alkaline) high side of the pH scale.

WHAT YOU'LL NEED

- A clean 2-cup (500-milliliter) measuring cup
- pH testing strips that can register from 6 to 11
- A fingertip-sized sample of base cleaner
- 2 ounces of soft, or mineral-free, water

Finding pH Strips

Finding and buying pH strips can be as easy as going to your local hardware or pool-supply store. Just be aware that some pH strips don't read between 6 and 11 pH, so call around first before you drive across town and waste a trip. It's best to do an Internet search to find the pH strips you need and get them online. Just type in "soap crafting pH test," and select the vendor you want.

Let's Start

1. Put about 2 ounces of soft water in a clean glass measuring cup.
2. Stir the bucket of base cleaner well to make sure that you get a uniform mix from which to pull a sample. Put about half a teaspoon of your base cleaner into the water in your measuring cup.
3. Stir until the base dissolves and mixes well; the water should look milky.
4. Get a pH strip, or if you have a roll of it, tear off a piece that is about 1½ inches long.
5. Dip the pH paper into the solution, and then immediately compare it with the color chart on the pH paper container.
6. That's all there is to it. Write down the pH number, and label the base container with that number. Now you have a pH reference for your batch of base.

TIP Be careful not to get your fingers in the solution you're testing. This can adversely affect the pH level.

Project 27
Making a Base Liquid pH-Balanced Laundry Soap

To formulate a laundry detergent with the base cleaner, you'll need to dilute it with soft, or mineral-free, water as well as balance its pH. For a detergent to work safety and not damage your laundry, you need to make it close to neutral on the pH scale. A neutral pH means that it's not so harsh as to damage your laundry or so weak that it doesn't remove dirt.

That's a pHact, Jack

For a good laundry detergent, you will have to have as neutral a pH as possible. Acidic and basic are two extremes that describe a chemical property. When an acid and a base are mixed, they cancel each other out and neutralize their extreme effects. A substance that is neither acidic nor basic is neutral. The pH scale measures how acidic or basic a substance is. The pH scale ranges from 0 to 14. A pH of 7 is in the middle of the pH scale, so it is considered neutral. Thus a pH less than 7 is acidic, and a pH greater than 7 is basic. You want your detergent to be no lower than 7 and no greater than 9 on the pH scale.

TIP Just 1 liter of this base-formulated liquid laundry soap can wash 24 average loads of laundry. At the store, this amount of detergent would cost over $13 for a name brand. Your cost with this formula can be less than $1 for 24 loads.

TIP This cleaner also can be used to remove tough stains. Pretreat the stained item by gently rubbing a small amount of cleaner onto the stain, and let it sit for a few minutes, not allowing it to dry. Then wash as usual.

WHAT YOU'LL NEED

- A 1-gallon (3.8-liter) plastic container with a lid (such as a milk jug)
- A 2-cup (500-milliliter) measuring cup
- pH testing strips that can register from 6 to 11
- 2 cups (500 milliliters) of base cleaner
- A bottle of lemon juice concentrate or white vinegar (pH of 2 to 4)
- 2 cups (500 milliliters) of soft, or mineral-free, water
- A box of borax or baking soda (pH of 8 to 9)

Let's Start

1. Make sure that your base cleaner is mixed well, and measure out 2 cups (250 milliliters) into the measuring cup. Now pour that into the 1-gallon (3.8-liter) sealable plastic container.
2. Measure out 2 cups (250 milliliters) of soft, or mineral-free, water, and add it to the plastic container. Put the lid on it, and give it a good shake. It's ready to use if your pH is at least 7 and no greater than 9; if not, continue on with the next few steps.
3. What is your base's pH number? If you don't know, follow the steps from Project 26. Knowing the pH is important to protect your laundry and to ensure that any dirt is removed.
4. If you have a pH above 9, you need to add an acid to bring it down. Use the lemon juice or white vinegar. If you have a pH lower than 7, you need to add a salt or base to bring it up. Use the borax or baking soda. Add (over a sink or outside in case of a spill) a tablespoon of what you need to make a neutral pH to the plastic container holding the water and base cleaner you just mixed. Carefully swirl the container around or mix it with the tablespoon.

TIP Don't put the lid on the container too soon so as to allow a small amount of carbon dioxide gas to escape. You can have a reaction similar to mixing baking soda and vinegar, like when you were a kid in science class.

5. Now squeeze the container to push some of the air out of it, and then put the lid on.
6. Put your thumb on the lid, and give the container a good shake. If it starts to build pressure, stop and open the lid. Keep repeating these steps until you think it's mixed well.
7. Now check the pH with a pH strip, or if you have a roll of pH paper, tear off a piece that is about a finger's length or so.
8. Carefully pour a small amount of the mixture into a clean measuring cup, and dip the pH paper into it. Immediately compare the paper with the color chart on the pH paper container.
9. Keep repeating steps 4 through 8 until the pH is at 7 or 9.
10. If you want to add a fragrance, use an essential oil such as lavender. Just remember, a little goes a long way.

11. Label the container so that you know what it contains and how to use it.

How to Use Your Laundry Soap

Add 2 ounces (60 milliliters) of your liquid laundry soap to normal-sized loads of laundry. Adjust accordingly for larger or smaller loads. Pretreat by gently rubbing a small amount onto any stains, and let the item sit for a few minutes, not allowing it to dry. Then wash as usual. Clothing should be tested for colorfastness before using any type of bleach, bleaching solution, or strong cleaning product.

CAUTION To test colorfastness, find a hidden seam of the garment or a hidden spot. Apply the cleaner to the garment, and then dab the area with a clean cotton cloth. If the color leaches from the garment onto the cloth, you should not use the cleaning product on that item.

Project 28
Making a Base Surface-Cleaner Spray

Surface cleaners are literally "part of wetter water" because they act to reduce the surface tension of a liquid, especially water. Being part hydrophilic means that it is attracted to water and soluble in water, usually because it has a positive or negative charge. The other part of the surface-cleaning agent is hydrophobic, meaning that it is repelled by water. Surface cleaners help us to remove grime, glop, and gunk by making them easy to rinse away or stick to a rag based on how they push or pull water to chemically pry things loose. Our surface cleaner will have a ratio of 50 parts water to 1 part base cleaner. Use this to clean the kitchen, bathroom, sinks, stoves, and garbage cans.

WHAT YOU'LL NEED

- An old empty and clean triggered spray bottle that holds about 24 fluid ounces (650 milliliters)
- A 2-cup (500-milliliter) measuring cup
- A tablespoon
- A funnel
- Soft, or mineral-free, water
- Base cleaner (before opening, give it a shake or swirl to mix it well) (Figure 8-21)

Figure 8-21

Let's Start

Open the clean triggered spray bottle, and with the funnel, add 2 tablespoons (30 milliliters) of base cleaner. Now add just enough soft, or mineral-free, water to almost fill the bottle. Try to leave some airspace. Close the bottle, and shake it well. Give it some pumps to prime it, and you are ready to clean. Label it so that you remember what it is and how to use it.

TIP If you have some citrus enzymes made up, add a tablespoon to the contents of the spray bottle to make it an even better cleaner.

How to Use Your Surface-Cleaner Spray

Just spray over the area you want to clean, and wipe, rub, or scrub the area with a damp cotton rag. For real tough grime, spray the cleaner on a "soft" bristle pad to dampen it, and scrub the gunk away. In hard-to-reach areas, use an old toothbrush to attack the grim. Use clean water to rinse away residue, or wipe it away with a dampened cotton rag.

Project 29
Making a Base Floor Cleaner

To clean a solid-surface floor, you first need to know the type of material the floor is made of and follow the floor manufacturer's recommendations. A floor can be easily damaged if you use the wrong method to clean it. The one thing all solid surfaces have in common is they can be swept. When you clean your floors, the first step to prevent your floor from becoming a sticky, muddy mess when you mop is to either sweep or vacuum it to remove any loose materials. You'll also want to remove any sticky or gunky spots that you notice when sweeping or vacuuming. Choose a mop based on your floor type. If you have a floor with a lot of texture, you'll want the more classic white string or rag mop. If you have a smooth floor, a sponge mop will work well.

Nonwooden Floors

When you wet mop nonwooden floors such as vinyl, ceramic, and slate, you should use a surface cleaner and a mop with a great deal of water. A wet mop is used to apply the water and surface cleaner to gather and remove the released dirt and any residual cleaner that could leave a film.

Wood, Laminant, and Marble Floors

Damp mopping is a method in which you wet the mop just slightly (wet it and wring it out almost dry). Go over the floor once to dissolve or loosen the dirt, and after rinsing the mop and ringing it, give the floor one more mop pass to remove the dirt. If you have a prefinished floor, it's a good idea to dry the floor with a towel or buffer.

WHAT YOU'LL NEED

- A vacuum cleaner or broom and dust pan
- A mop
- Two mop buckets
- A 2-cup (500-milliliter) measuring cup
- A tablespoon
- Soft, or mineral-free, water
- Base cleaner (before opening, give it a shake or swirl to mix it well)
- White vinegar (Figure 8-22)

Figure 8-22

Let's Start

1. Use a bucket in which to mix the cleaner and water because it can be carried along the path of the mop.
2. Pour a gallon of soft, or mineral-free, water into the bucket, and add 4 tablespoons (60 milliliters) of the base cleaning solution and ½ cup (120 milliliters) of white vinegar.
3. Fill another mop bucket with fresh water with which to rinse out the mop. You also can do this in your sink if you want. If you use soft, or mineral-free, water to rinse, you may find your floor looking cleaner because the soft water will not leave minerals behind that can cause a film.
4. Dip your mop in the bucket, and wring it out with a wringer or by hand. Too much water dripping from the mop can damage the floor or leave the floor needing more drying time.
5. Mop from the top of the room to the bottom so that you are always standing in an area that hasn't been mopped yet. This will prevent you from tracking footprints or slipping as you work. If you are using a damp sponge mop, try to mop in straight lines. For rag mops, move the mop in a figure-of-eight shape, and let it flop over itself to use the design of the rag mop effectively. When you encounter a tough or sticky spot, rub it quickly, and press down with the mop to scrub it away.
6. Rinse the mop frequently in the other bucket or your sink, and wring the rinse water out. Wet the mop in your cleaning solution bucket again, and begin mopping. When the cleaning solution or rinse water becomes dirty, it needs to be changed. You can't clean dry dirt with wet dirt.
7. When you've mopped every section, rinse the mop again thoroughly, rinse out the mop bucket, and allow to dry completely so that it doesn't get stinky. Let your floor dry completely before walking on it.
8. Label the cleaning solution so that you remember what it is and how to use it.

TIP **Do not think that extra mopping solution will get you a cleaner floor. It will just leave a residue on the floor. If you use too much cleaning solution, add more water to the bucket to reduce the cleaning solution. Remember to add less cleaning solution next time you mop.**

TIP **Flooring types are varied and may need modified cleaning care. Know what type of floor you have, and follow the manufacturer's directions.**

TIP **If you have some citrus enzymes made up, add a tablespoon to the mop bucket to make the solution an even better cleaner.**

Project 30
Making a Base Carpet Shampoo

Getting a professional to clean your carpets is expensive. Many of us own carpet shampooers and are able to clean our carpets more frequently than people who have to hire someone to do the job. The manufacturers of carpet cleaning machines recommend using only their products to clean your floors, and those products are often overpriced. When your carpet does get soiled, it is important to remove as much of the matter immediately. For wet messes, you should blot the carpet dry with paper towels as soon as possible. Dry messes should be picked up or swept or vacuumed up immediately. Once you have picked up the mess as much as possible, you now can use a cleaning solution and your shampooer.

WHAT YOU'LL NEED

- A vacuum cleaner
- A 2-cup (500-milliliter) measuring cup
- A tablespoon
- An empty and clean 68-fluid-ounce (2-liter) bottle (such as a soda bottle)
- A funnel
- Soft or, mineral-free, water (tap water is optional for carpet shampooing)
- Base cleaner (before opening, give it a shake or swirl to mix it well)
- White vinegar
- Fragrance oil (optional) (Figure 8-23)

Figure 5-23

Let's Start

When you clean your carpets, the first step is to prevent your floor from becoming a bigger mess when you shampoo it, so you must vacuum the floor to remove any loose particles. You'll also want to remove "by hand" any stained or extra dirty spots that you notice when vacuuming. Make sure that your carpet shampooer is clean and free of dirty water or old cleaner that was left in it from last time it was used.

TIP **Carpeting types are varied and may need modified cleaning care. Know what type of carpet you have, and follow the manufacturer's directions.**

1. Open the 68-fluid-ounce (2-liter) bottle, and with your funnel, add 7 cups (1,750 milliliters) of water.
2. Add 2 tablespoons (30 milliliters) of the base cleaner to the water in the bottle.
3. Fill the bottle the rest of the way with white vinegar.
4. Add a few drops of a fragrance oil if you would like. Just use it sparingly because a little goes a long way.
5. Label the cleaner so that you remember what it is and how to use it.

How to Clean with the Shampooer

1. Remove the furniture from the room before starting the cleaning process.
2. Then vacuum the carpet thoroughly.
3. Now apply the mechanical brush on the area. This brush will work on the filthy portions by injecting the foamy detergent solution into the carpet.
4. When completing the cleaning process, allow the carpet to dry completely. You will have to wait overnight before walking on it.
5. Vacuum the carpet once again to collect any loosened fibers. Move your furniture back in, and you are done.

TIP **Follow the carpet and machine manufacturers' instructions when using the shampooer.**

TIP **If you have some citrus enzymes made up, add a tablespoon to the bottle of solution to make it an even better cleaner.**

Project 31
Making a Base Carpet Spot and Stain Remover

The most important rule for cleaning any carpet stain is to clean the spilled matter immediately. If you don't act right away, you may be ruining your carpet. If you spill something on your carpet, grab a cloth. First, clean up any solid matter, and then blot (up and down) at the stain until you've lifted as much of the stain as possible.

If a considerable amount of liquid has been spilled, weeping may occur. If you've ever cleaned a stain and had it reappear a day or two later, your carpet is suffering from weeping. This means that the liquid has pooled under the carpet. Even though you blotted up the initial stain, you only cleaned the surface. Eventually, the liquid works its way back up the fibers to the top of the carpet, causing it to look like the stain has reappeared. To prevent weeping, cover the area with a thick cloth, and weight it down with books. Leave overnight, and then remove the stain normally. When it's clear that no more of the stain will come out, it's time to bring out your carpet cleaner.

WHAT YOU'LL NEED

- An empty and clean triggered spray bottle that holds about 24 fluid ounces (650 milliliters)
- A 2-cup (500-milliliter) measuring cup
- A tablespoon
- A funnel
- A 2½-fluid-ounce (325-milliliter) brown plastic bottle of 3% hydrogen peroxide (like that sold for wound disinfectant)
- Base cleaner (before opening, give it a shake or swirl to mix it well)
- Borax or baking soda (Figure 8-24)

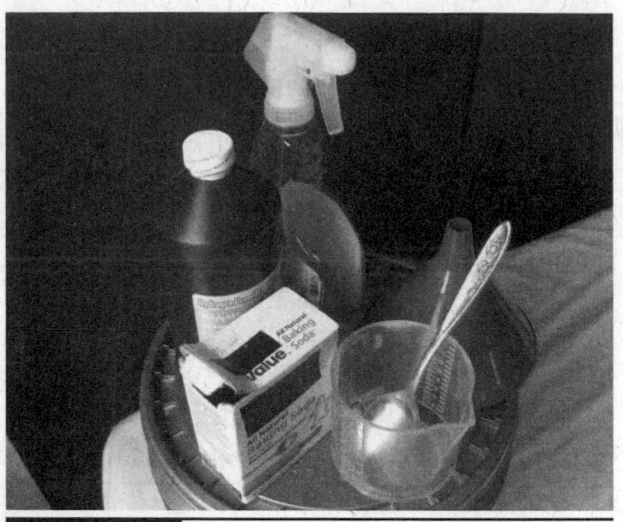

Figure 8-24

Let's Start

1. With the funnel, add 1 tablespoon (15 milliliters) of borax or baking soda to the clean triggered spray bottle.
2. Also add 1 tablespoon (15 milliliters) of base cleaner to the spray bottle.
3. Again using the funnel, add just enough 3% hydrogen peroxide to almost fill the bottle. Try to leave some airspace.
4. Close the bottle, and shake it well. Give the bottle some pumps to prime it, and you are ready to clean.
5. Label the bottle so that you remember what it is and how to use it.

TIP If you have some citrus enzymes made up, add a tablespoon to the spray bottle to make the solution an even better cleaner.

TIP Don't scrub the stain. This will only weaken the carpet fibers and cause the stain to spread. Blotting gently will remove the stain without causing additional damage.

How to Use Your Carpet Spot and Stain Remover

Always be sure to test the carpet for colorfastness before using any stain remover solution. Test in an inconspicuous spot such as the back corner of a closet. Always use a blotting (up and down) motion to remove any stain, working from the outside in to prevent spreading the stain. If the stain is dried or old, scrape or brush off any crusted matter, and vacuum it away.

1. Spray cleaner onto the spot.
2. Allow it to penetrate the stain for several minutes.
3. Blot the stain gently with a clean white cotton towel or rag.
4. Gently rub with another clean towel or rag and clean water.
5. Repeat steps 1 through 4 if necessary.
6. Allow the carpet to dry before resuming use.

Project 32
Making a Base Glass Cleaner

WHAT YOU'LL NEED

- An empty and clean triggered spray bottle that holds about 24 fluid ounces (650 milliliters)
- A 2-cup (500 milliliter) measuring cup
- A tablespoon
- A funnel
- Soft, or mineral-free, water
- Base cleaner (before opening, give it a shake or swirl to mix it well)
- White vinegar (Figure 8-25)

Figure 8-25

Let's Start

1. With the funnel, add 1 tablespoon (15 milliliters) of base cleaner to the triggered spray bottle.
2. Again with the funnel, add 2 cups (500 milliliters) of soft, or mineral-free, water to the bottle.
3. Pour in the white vinegar to almost completely fill the bottle, but try to leave some airspace.
4. Close the bottle, and shake it well. Give the bottle some pumps to prime it, and you are ready to clean.
5. Label the bottle so that you remember what it is and how to use it.

TIP If you have some citrus enzymes made up, add a tablespoon to the spray bottle to make the solution an even better cleaner.

How to Use Your Base Glass Cleaner

Windows look best if they are cleaned on a regular basis—at least twice a year on the inside and outside. Spray the cleaner on the glass, and wipe away both the dirt and cleaner with paper towels, a clean cloth, or newspaper.

In the Newspaper

Newspapers are good for cleaning glass because the paper leaves very little lint. Compared with paper towels, newsprint is much more rigid, and hence the fibers will not separate individually like they do with a paper towel. This is the source of the lint. The ink does not come off for two reasons. First, you are wiping a mirror or glass, which is a highly polished surface, so there is nothing for the dried ink to stick to. Second, on wetting the newsprint with liquid, the ink becomes infused in the fibers of the newsprint. Ink is nothing more than a stain. It is hard to stain windows with ink yet easy to stain paper. Glass does not absorb ink, whereas paper will. And all the dirt on the window or mirror also will be absorbed into the newsprint.

Window Washing Rules

1. Wash one side of a window with horizontal strokes and the other side with vertical strokes so that you can pinpoint which side of the window has a streak.
2. Use a squeegee on a long handle or a sponge-squeegee combination to prevent streaks on large windows.
3. Washing windows should be done on cloudy days because direct sunlight dries cleaning solutions before you can polish the glass properly.
4. Wash windows from top to bottom.

Project 33
Making a Base Disinfectant Surface Cleaner

Disinfectants kill bacteria on the surfaces of nonliving objects. Disinfectants are used extensively in hospitals and other medical treatment centers, where they help to maintain a clean environment to prevent the spread of bacteria from objects to people. Disinfectants used properly are still able to kill bacterial strains that have become resistant to antibiotic treatments.

In the last 20 years, the use of disinfectants such as antibacterial sprays has increased in the home, but this may not necessarily be a good thing. Most bacteria in our home environments are friendly bacteria and do not cause disease. Using antibacterial products all over the home probably is unnecessary. Cleaning danger areas such as toilets, drains, and rubbish bins with a suitable disinfectant or bleach solution is very likely sufficient. Use disinfectants only to clean the kitchen, bathrooms, sinks, stoves, and garbage cans.

WHAT YOU'LL NEED

- An empty and clean triggered spray bottle that holds about 24 fluid ounces (650 milliliters)
- A 2-cup (500-milliliter) measuring cup
- A tablespoon
- A funnel
- A 34-fluid-ounce (1-liter) bottle of cheap vodka (70 to 100 proof alcohol content)
- Base cleaner (before opening, give it a shake or swirl to mix it well) (Figure 8-26)

Figure 8-26

Let's Start

1. Using the funnel, add 2 tablespoons (30 milliliters) of base cleaner to the spray bottle.
2. Again using the funnel, add just enough of the vodka to almost fill the bottle. Try to leave some airspace.
3. Close the spray bottle, and shake it well. Give the bottle some pumps to prime it, and you are ready to disinfect and clean.
4. Label the bottle so that you remember what it is and how to use it.

TIP **If you have some citrus enzymes made up, add a tablespoon to the spray bottle to make the solution an even better cleaner.**

How to Use Your Disinfectant Surface Cleaner

1. Just spray mist over the area you want to clean, let it set for up to 10 minutes to kill germs, and then wipe, rub, or scrub the area with a damp cotton rag.
2. Use clean water to rinse away residue, or wipe it away with a damp cotton rag.

Project 34
Making a Base Abrasive Toilet Bowl Cleaner

Cleaning a toilet can be a disgusting task, but it's something that has to be done. The right way prevents the spread of bacteria and viruses. The greener way of cleaning a toilet is safer and saves time and energy. Prevent extra cleanup by removing all excess items from around the toilet. Don't forget to remove anything on top of the tank to prevent dropping items into the bowl during cleaning. Do not use sponges when you scrub a toilet because they are a great place to breed bacteria. Use reusable cloths or rags, and wash them separate from other laundry as soon as you are done.

WHAT YOU'LL NEED

- A clean, empty 68-fluid-ounce (2-liter) bottle (such as a soda bottle)
- A 2-cup (500-milliliter) measuring cup
- A tablespoon
- A pair of comfortable scissors
- A toilet brush
- Gloves
- Antibacterial cleanser for the outside of the toilet
- Cotton rags, towels, or cloths
- A 1-pound (454-gram) box of baking soda
- Base cleaner (before opening, give it a shake or swirl to mix it well) (Figure 8-27)

TIP **Cut off the top third of an empty 68-fluid-ounce (2-liter) bottle, and the bottom two thirds of the bottle makes a perfect rounded container in which to put a toilet brush.**

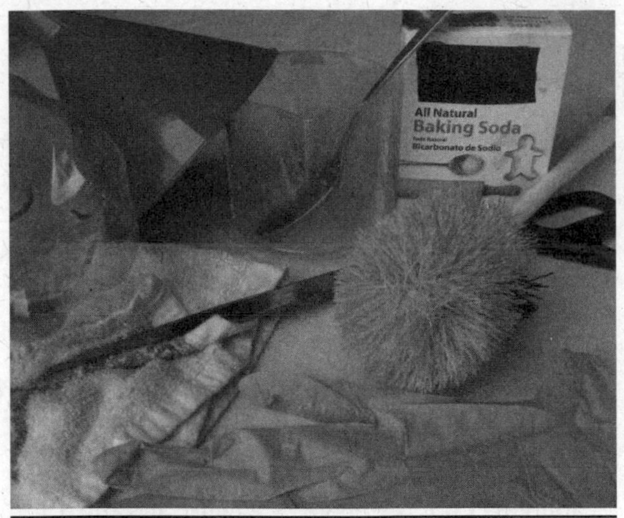

Figure 8-27

Let's Start

1. Cut the top third off of an empty 68-fluid-ounce (2-liter) bottle (save the top of the bottle to use as a funnel).
2. Add ½ cup (125 milliliters) of baking soda to the bottom two-thirds of the cut bottle.
3. Add 2 tablespoons (30 milliliters) of base cleaner to the baking soda in the bottle bottom. Use the tablespoon to mix it into a paste that isn't runny so that it sticks to your toilet brush (if it's too thick to use, add a little water until you're happy with the texture).
4. Now, with your gloves on (if you want), insert the brush into the cleaning paste you just made, coating the brush.
5. Now you can wipe and scrub the toilet bowl to clean it.

TIP If you want to make this in bulk, just use a resealable old butter dish or a large glass jar with an airtight lid, and add to it 1 cup (250 milliliters) of baking soda mixed with 4 tablespoons (60 milliliters) of base cleaner, and mix it well. Remember to label it so that you know what it is and how to use it.

How to Clean the Inside of the Toilet Bowl

1. Begin cleaning the bowl from the top down.
2. Always begin scrubbing under the rim first.
3. Dedicate plenty of scrubbing time under the rim to get to all the stains and grime.
4. Next, scrub the bowl.
5. Finally, scrub the hole at the bottom of the toilet.
6. Flush the toilet with the lid down.
7. Wipe up any drips or spills.

TIP If you have some citrus enzymes made up, add a tablespoon to the bottle to make the solution an even better cleaner.

Project 35
Making a Base Rust and Mineral Stain Remover

Removing rust and mineral stains is impossible in situations where chemical changes to the materials have occurred. If this formula doesn't work, try a store-bought cleaner with oxalic acid, which is effective as a rust stain and mineral remover. Oxalic acid is made commercially by treating carbohydrates with nitric acid or cellulose with sodium hydroxide. It's not what I would call a green or natural cleaner, but it beats replacing a perfectly good working toilet with a new one because it looks bad. Fixing or reusing something is adding value and keeping it from being wasted. Saving something from being wasted is also green.

WHAT YOU'LL NEED

- An empty and clean 68-fluid-ounce (2-liter) bottle (such as a soda bottle)
- A 2-cup (500-milliliter measuring cup
- A tablespoon
- A funnel
- Base cleaner (before opening, give it a shake or swirl to mix it well)
- White vinegar (Figure 8-28)

Figure 8-28

Let's Start

1. With the funnel, add 2 tablespoons (30 milliliters) of base cleaner to the 68-fluid-ounce (2-liter) bottle.
2. Then fill the bottle the rest of the way with white vinegar, and replace the lid.
3. Give it a good shake.
4. Remember to label the bottle so that you know what it is and how to use it.

How to Use It

Solid Surfaces

1. Add the mix to a recycled clean and empty spray bottle.
2. Spray a mist over the area you want to clean, and let it sit a few minutes.
3. Then you can wipe, rub, or scrub the area with a damp cotton rag. For tough deposits, spray the cleaner on a soft bristle pad to scrub the gunk away. In smaller, hard-to-reach areas, use an old toothbrush to reach the grim.
4. Use clean water to rinse away residue, or wipe it away with a damp cotton rag.

The Toilet

Because it is necessary to reduce the amount of water left in the bowl, you should do the following:

1. Push the toilet brush in and out of the trap quickly until the water level drops in the trap.
2. If that doesn't work, use a toilet plunger to force the water down and out of the trap.
3. Once you have gotten out as much water as you could, slowly pour about 1 cup (250 milliliters) of cleaner into the bowl.
4. Move the brush around to scrub out the stains. Continue to do this on and off for a few minutes.

TIP **You'll see a difference soon. If not, your stains are either set in or etched into the surface, in which case you may want to try a more powerful store-bought cleaner. Please remember to read and follow the manufacturer's directions.**

Project 36
Making a Base Mold and Mildew Remover

Mold and mildew cause considerable damage if permitted to grow, in addition to causing an unpleasant musty odor. They need moisture, darkness or limited light, little air movement, and food. Areas prone to their effects are basements, crawlspaces of houses, and closets. Mold and mildew can grow rapidly on anything from which they can get food, particularly cellulose products such as cotton, linen, wood, and paper, as well as protein substances such as silk, leather, and wool. Molds are simple plants and belong to the group known as fungi. Although molds are always present in the air, those that cause mildew on solid surfaces such as shower walls only need moisture and a certain temperature to grow.

Prevention Is Best

Prevention is the best mildew policy. If things are kept clean, well ventilated, and dry, your chances of having mildew are greatly lessened.

WHAT YOU'LL NEED

- An empty and clean resealable butter container or a large glass jar with an airtight lid big enough to hold about 4 cups (1,000 milliliters)
- A 2-cup (500-milliliter) measuring cup
- A tablespoon
- Two 12-fluid-ounce (325-milliliter) brown plastic bottles of 3% hydrogen peroxide (like those sold for wound disinfectant)
- Base cleaner (before opening, give it a shake or swirl to mix it well)
- Borax (Figure 8-29)

Figure 8-29

Let's Start

1. Add 1 cup (250 milliliters) of borax to the resealable butter container or large-mouthed glass jar with an airtight lid.
2. Mix in another 4 tablespoons (60 milliliters) of base cleaner.
3. Follow that with 12 fluid ounces (325 milliliters) of hydrogen peroxide.
4. Mix it all together with the tablespoon until the solution is uniform and pourable (if it's too thick to pour, add a little water until you're happy with the texture).
5. Remember to label the container so that you know what it is and how to use it.

How to Use It

1. Dab a moistened rag into your mold and mildew cleaner. Then rub it over the area you want to clean, let it sit a few minutes, and then wipe, rub, or scrub the area with a damp cotton rag.
2. For real tough mildew stains, use the cleaner with a soft bristle pad to scrub the gunk away. In hard-to-reach smaller areas, use an old toothbrush to reach the grime.

3. Use clean water to rinse away residue, or wipe it away with a damp cotton rag.

TIP Mildew stains sometimes have grown into the surface and can't be removed.

TIP If you have some citrus enzymes made up, add a tablespoon to the bottle to make the solution an even better cleaner.

Project 37
Making a Base Oven Cleaner

The hardest part of cleaning an oven is getting all that baked-on black stuff off the walls of the oven. If you have a self-cleaning oven, it basically uses high heat to cremate the stuff into an easy to wipe away ash. That crusty stuff is basically burnt sugars and starches, as well as polymerized fatty acids from spilled oil and fats. Chemical oven cleaners work by breaking that stuff stuck on the walls of the oven down into smaller particles and soaps that you can easily wipe out with some soapy water.

WHAT YOU'LL NEED

- An empty and clean resealable butter container or large glass jar with an airtight lid big enough to hold about 4 cups (1,000 milliliters)
- A 2-cup (500-milliliter) measuring cup
- A tablespoon
- Cotton rags
- Base cleaner (before opening, give it a shake or swirl to mix it well)
- White vinegar
- Baking soda (do not add to the mix) (Figure 8-30)

Figure 8-30

CAUTION Before using anything to clean your oven, get out your Use and Care Guide (or call the manufacturer for a new one). There are several types of ovens, and you could ruin your oven by using cleaners that are wrong for your type of oven. After learning what kind of oven you have, you can determine the best way to clean it.

Let's Start

1. Add 2 cups (500 milliliters) of white vinegar to the recycled butter container or large glass jar.
2. Now add 2 tablespoons (30 milliliters) of the base cleaner.
3. Mix it with the tablespoon.
4. Label the container so that you remember what it is and how to use it.

How to Use It

1. Make sure that your oven is off and cool and that you won't need to use it for 24 hours.
2. Wet the cotton rags with the cleaner. Do this inside the oven so that you don't drip cleaner everywhere with the rags.

3. Lay the wet rags flat on the spots you want to clean. Make sure that they are soaking wet with the cleaner.
4. Let the wet rags sit on those spots overnight or longer.
5. Remove the rags, and then sprinkle a thin layer of baking soda over the area you want to clean.
6. Now, with a clean damp rag, scrub away the residue.
7. Rinse the rags often until your oven is clean.

TIP **Don't let the rags dry out while they are sitting on the spills or they won't do any good at all.**

TIP **If you have some citrus enzymes made up, add a tablespoon to the cleaner to make the solution an even better cleaner.**

Project 38
Make Your Own Supercleaning Citrus Fruit Enzymes

Enzymes are proteins produced by all living organisms that as far as cleaning goes act as a catalyst to push along chemical reactions that occur at slow rates or not at all. Enzyme catalysts help reactions to move from beginning to end. Enzymes often stay intact or are not "exhausted" in cleaning reactions, so they are available to help in multiple reactions. Thus a little goes a long way.

As with any type of catalyst, an enzyme facilitates the chemical reaction without being used up or destroyed, leaving the enzyme available for yet another reaction. Eventually, all the reactions are complete and the enzyme stops working. Over time, the enzyme will biodegrade on its own.

The enzymes used for cleaning act on the substrates of the materials that make up a variety of stains and soils and break down their chemical bonds so that they can be washed away more easily. Since one enzyme molecule can act on bio-based grime, gunk, and glop, just a small amount of enzyme added to a cleaner can make a big difference. Best of all, enzymes can be made from natural things such as the waste we have in our homes every day.

Fruit Enzymes

Fruit ripens because of enzymes, and those enzymes are created because of a gas called *ethylene* that is a product of fermentation. This process is a natural one caused by age and natural decomposition or damaged fruit. We've all heard the old phrase, "One bad apple spoils the bunch." Why? Because overripened, wounded, or a worm-infested apple produces larger amounts of ethylene than a good apple. One bad apple can cause good apples to ripen prematurely. Apples that are too ripe are bruised by their own weight and become soft, mushy, and gross. This occurs because the good apples' cells respond to the greater amounts of ethylene produced by the damaged fruit's enzymes as a result of escalated fermentation and decomposition, which causes the good apples to make more enzymes and more ethylene gas, and the whole batch is spoiled. Ethylene apparently "turns on" the genes that are responsible for producing these enzymes in the fruit's cells. The enzymes then catalyze reactions to alter the characteristics of the fruit to ripen it quickly. It's a matter of survival for plants to have ripe fruit so that animals and insects can help to spread their seeds, so don't feel sorry for the good apples gone bad.

Does This Smell Bad to You?

The action of the enzymes causes the ripening response. Chlorophyll is broken down, and sometimes new pigments are made, so the fruit skin changes color from green to red, yellow, or blue. Acids are broken down, so the fruit changes from sour to neutral. The degradation of starches produces sugar and increases the fruit's juiciness. The breakdown of pectin between the cells of the fruit releases bonds so that the pectin fibers can slip past each other, resulting in a softer fruit texture. For odors, enzymes break down large organic molecules into smaller ones that can become volatile and evaporate into the air, producing an aroma that we and other living creatures can detect.

How This Enzyme Recipe Works (From My Perspective, Anyway)

The ingredients we're using are from the fermented peels of citrus fruits, natural sugars, and water. The cells and compounds in our ingredients are full of enzymes, but not enough yet for us to use as a cleaner. We need to multiply or make copies of those natural enzymes to increase the enzyme concentrations. Thus we are going to employ living microorganisms to do just that via a process called *fermentation*. Fermentation is the anaerobic conversion of sugars to carbon dioxide and alcohol induced by "non-oxygen-loving" microbes that split complex organic compounds into relatively simple substances and produce our enzymes. The alcohol produced by fermentation is called *ethanol*, like the alcohol found in wine, beer, and liquors, but I wouldn't recommend that anyone drink our cleaner. The ethanol produced by this fermentation is beneficial in helping to keep in check competing microorganisms such as black mold and fungi that we don't want or need for proper enzyme production, and it also adds a solvent to our cleaner. The fermentation process speeds up the production and reproduction of the enzymes we want to use for cleaning.

WHAT YOU'LL NEED

- An electric blender or food processor
- A 2-cup (500-milliliter) measuring cup
- A clean, empty 68-fluid-ounce (2-liter) bottle (such as a soda bottle)
- A funnel
- A large balloon (such as a punch-ball balloon)
- A piece of string or a twist tie (such as from a loaf of bread)
- About 4 cups (1,000 milliliters) of soft, or mineral-free, water
- At least 3 cups (750 milliliters) of chopped citrus skins from any combination of oranges, lemons, grapefruit, and limes
- 1 cup (250 milliliters) of natural sugars only (natural brown sugars, honey, and molasses are okay)
- 2 tablespoons (30 milliliters) of active dry yeast (such as that sold for making bread) (Figure 8-31)

Figure 8-31

TIP You can freeze the citrus peels until you have enough to make your cleaner.

Let's Start

1. With a funnel, pour in 1 cup (250 milliliters) of natural sugar, 2 tablespoons (30 milliliters) of yeast, and 2 cups (500 milliliters) of water into the 68-fluid-ounce (2-liter) bottle. Put the lid on, and shake well until the sugar dissolves.
2. Now cut up the citrus waste, and process it in the blender or chop it by hand until you have a pourable consistency and enough of it to make 3 cups (750 milliliters) (Figure 8-32).
3. Again using the funnel, put the 3 cups (750 milliliters) of finely chopped citrus into the bottle. Put the lid on, and shake it well.
4. Now remove the lid, and again using the funnel, add the remaining 2 cups (500 milliliters) of water to the bottle (leave a few inches of airspace at the top). Replace the lid once more, and shake it again (Figure 8-33).
5. Remove the cap, and roll the balloon over the top of the bottle's neck. Use a twist tie or string to wrap around the surface of the balloon that is in contact with the bottle's neck so that the balloon can't be removed easily. The balloon captures and holds the CO_2 gas that is a by-product of the fermentation process and keeps other microbes out (Figure 8-34).
6. For the first month, you need to gently swirl the mixture almost daily. Don't worry if you see a grayish moldy looking layer form on top of the liquid. This is just the good bacteria doing its job and is not mold. The balloon will begin to fill, but it'll deflate by the end of the process. If your balloon bursts, just put another one on.

Figure 8-33

Figure 8-32

Figure 8-34

Figure 8-35

7. After 2 weeks, the enzymes should be ready to use. The color should be brown. If the mixture turns black, add more sugar, replace the balloon, and let it ferment a little longer (Figure 8-35).
8. Remember to label the container so that you know what it is and how to use it.

How to Use Your Fruit Enzyme as a General Cleaner

Add 2 tablespoons (30 milliliters) per liter of water for a household cleaning liquid to remove foul odors and for mopping floors.

How to Use Your Fruit Enzyme Cleaner as a Laundry Detergent Booster

Add 2 tablespoons (30 milliliters) per load of laundry.

How to Use Your Fruit Enzyme Cleaner as a Super Stain Lifter

1. Dab water on the stain with a wet towel.
2. Dab the enzyme solution on the stain, and let it sit for a few minutes.
3. Lightly scrub and rinse or wipe away with clean water or a clean wet rag.

CAUTION **Remember to always test for colorfastness. Find a hidden or inconspicuous spot, apply the enzyme to the material, and then dab the area with a moist clean cotton cloth. If the color fades or leaches from the material onto the cloth, you should not use the enzyme on that material.**

Project 39
Make Your Own Citrus Oil Extract

Everything that is in nature has multiple functions, and in the case of oranges, the skin has far more uses than just being a protective coating. The oil in the skin is called *d-limolene*. This is the chemical name for citrus oils. These oils can be used to replace industrial solvents and smell like citrus. Sometimes pure orange oil or orange essence is added for fragrance. A product can be called *orange oil* if it contains as little as 2% pure orange oil.

Because this process uses solvents, wear gloves, and keep out of the reach of children and away from open flames, and the same applies for the finished product. Orange oil is flammable and very corrosive. For most cleaning purposes, ¼ ounce (7 milliliters) mixed with 1 quart (1 liter) of water should be sufficient. Always spot test a brew before applying in quantity.

WHAT YOU'LL NEED

- An electric blender or food processor
- A 2-cup (500-milliliter) measuring cup
- A couple clean, empty 68-fluid-ounce (2-liter) bottles (such as soda bottles)
- A funnel
- An old cotton sock that's clean and free of holes or cheesecloth
- An empty and clean resealable butter container or large glass jar with an airtight lid that holds about 2 cups (500 milliliters)
- 3 cups (750 milliliters) of chopped citrus skins from any combination of oranges, lemons, grapefruit, and limes
- 25 fluid ounces (750 milliliters) of consumable 190 proof grain alcohol (Figure 8-36)

Figure 8-36

CAUTION 190 proof grain alcohol is extremely flammable, so be careful, look out for ignition sources, and use it in a well-ventilated space.

Let's Start

1. Dry the citrus peels. You can do this by placing them in a single layer in a warm dry place, preferably in sunlight. If you are in a hurry, you can dry them in your oven at its lowest heat setting. Bake the peels for several hours, testing them occasionally to see how dry they are (Figure 8-37).
2. Once you have enough dried peels to make 3 cups (750 milliliters) of ground peels (about 8 cups of dried ungrounded peels), add them to the blinder or food processor and give them a good grinding.
3. With your funnel, add the 3 cups (750 milliliters) of ground citrus peels to the clean 68-fluid-ounce (2-liter) bottle.
4. Add just enough of the 190 proof grain alcohol on top of the ground peels to completely cover them (Figure 8-38).
5. Put the lid on the bottle, and give it a good shake. Then let it sit overnight (Figure 8-39).
6. Repeat the shaking, and let it sit for three or four days. Remember to carefully open the bottle to let any gases escape from time to time.

Figure 8-37

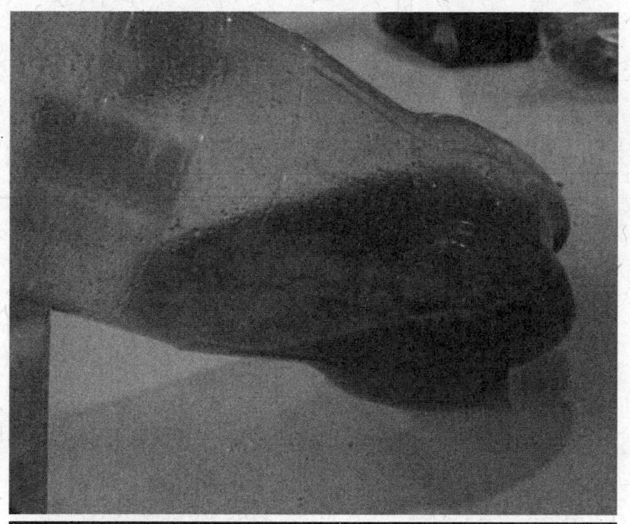

Figure 8-38

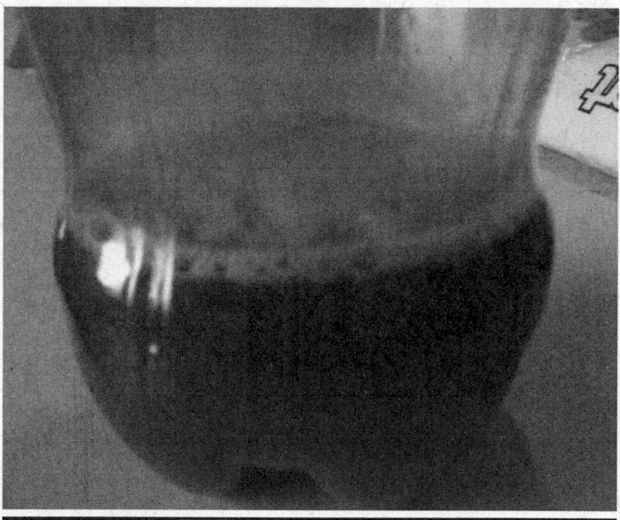

Figure 8-39

7. Set up a second open and empty 68-fluid-ounce (2-liter) bottle with the funnel in it.
8. Now hold that old cotton sock that's clean and hole-free over the funnel, and pour the contents of the first bottle into the sock "filter" and let the liquid that passes through the sock go into the second bottle.
9. Once the first bottle is empty and the liquid has stopped dripping from the sock into the empty bottle, place the sock full of the ground citrus peels in a recycled butter container or large glass jar, put an airtight lid on it, and set it aside for use in Project 41.
10. Remember to label the container so that you know what it is and how to use it.

How to Use It

Use the citrus oil extract cleaner as a powerful degreaser on oil-based stains:

1. Work the oil extract cleaner into the stain, and then work a full-strength liquid detergent into the stain.
2. Wash the garment normally, with the recommended amount of detergent for a regular laundry load, and rinse.
3. Repeat the preceding as many times as needed if removal is incomplete the first time.
4. Remember to inspect for oil stains in bright light so that you can see them easily, especially before drying (machine drying can thermoset stains in).

Project 40
Make Your Own Plant Oil-Based Heavy-Duty Solvent

Heavy-duty solvents are used to remove hard-to-clean substances from solid surfaces, such as bugs and tar that get on your car from time to time. Most of the chemicals used for this task are petroleum-based, have high amounts of volatile organic compounds (VOCs), and are not very user-friendly. The action of the solvent that we are going to make helps to release the bond the guts and grime have with a smooth surface so that we can wipe them away with a clean rag.

WHAT YOU'LL NEED

- An empty and clean 34-fluid-ounce (1-liter) bottle (such as a soda bottle)
- A 2-cup (500 milliliter) measuring cup
- A tablespoon
- Cotton rags
- A funnel
- Base cleaner (before opening, give it a shake or swirl to mix it well)
- Citrus oil extract cleaner you made in Project 39
- Castor oil (this can be found online or in drug and health food stores)

NOTE Vegetable can be substituted for castor oil, but it won't work as well.

Why Castor Oil?

My personal favorite oil to use for making a cleaner is castor oil. Castor bean oil can be found online and in drug stores, health food stores, and hobby shops that sell model aircraft fuel. Castor oil is the *only* natural oil that has a polar bond to organic solvents such as alcohol, so it naturally is just a better choice for making a cleaner that has great flexibility and high quality.

Castor Oil

From the 1850s until the early 1970s, U.S. farmers produced castor bean crops in the central part of the United States, supplying beans to over 23 crushing mills. In the 1980s, castor bean crops all but disappeared in the United States, and today, castor bean production is urgently needed in the United States for several important reasons. There are hundreds of products derived from castor bean oil because of its unique qualities. Everything from medicines, to biofuels, to cosmetics, to protective coatings, to jet engine lubricants is made from castor oil, and some of them are deemed essential for our national defense by the Agricultural Materials Act of 1984. Currently, the United States is the largest importer of castor oil in the world, and we depend on foreign sources for all our supply. Our primary sources for this vital product are China, India, and, to a lesser extent, Brazil.

The castor bean is a unique agricultural commodity. The castor bean contains as much as 50 to 60 percent usable oils. This allows for production potential of up to 1,000 pounds of oil per acre.

Let's Start

1. With the funnel, add ½ cup (125 milliliters) of citrus oil extract to the clean 34-fluid-ounce (1-liter) bottle.
2. Now add 2 tablespoons (30 milliliters) of base cleaner to the bottle. Top the rest of the bottle off with about 1½ cups (375 milliliters) of castor oil or vegetable oil. Even if you add just a small amount of castor oil, this will help to keep the vegetable oil in suspension and make it clean better.
3. Put the lid on the bottle, and give it a good shake.
4. Remember to label the container so that you know what it is and how to use it.

How to Use It

To remove bug guts, tar, and tree sap from your car:

1. First, remove any grit or dirt that might scratch the paint.
2. Wash your car and let it dry in the shade so that the surface of it is cool.

3. Shake the closed bottle of your solvent well to be sure that it's mixed well.
4. Take a clean rag and hold it tightly to the top of the open bottle and keep it there as you carefully turn the bottle over to let the solvent soak that part of the rag.
5. Flip the bottle back over, and set it down on a level surface so that it doesn't spill. Better yet, put the lid back on it and hold the bottle in your hand.
6. Now take the part of the rag that's moistened with the solvent and dab it on the spots you want to remove, and with a swirling action, lightly scrub on the spot until it disappears. If your rag dries out, rewet it with the solvent. Repeat the steps on each spot.

TIP If you can find a lid that fits your bottle that flips open or has a pull-type top like the one used on water or sport drink bottles, the bottle will be easier to open and close and less likely to spill.

Project 41
Make Your Own Heavy-Duty Hand Cleaner

Working on things that use grease or oil or have grease and oil on them can leave a nasty mess on our hands that even soap and water hardly can touch. Thus, if you are using a greasy mower, working on a car engine, or cleaning a dirty BBQ grill, you will need a safe and quick hand cleaner to get life back to normal.

WHAT YOU'LL NEED

- A tablespoon
- Cotton rags
- Base cleaner (before opening, give it a shake or swirl to mix it well)
- Baking soda
- Castor oil or vegetable oil
- The old butter container or large glass jar with airtight lid in which you stored the ground citrus peels from Project 39

Let's Start

1. Open the container of ground citrus peels (remove the citrus grounds from the sock if you haven't yet).
2. Add 1 cup (250 milliliters) of baking soda to the container with the ground citrus peels.
3. Add 1 tablespoon (15 milliliters) of base cleaner.
4. Add 4 tablespoons (60 milliliters) of castor oil or vegetable oil to finish off your ingredients.
5. Mix it as best you can, and put the lid back on it.
6. Remember to label the container so that you know what it is and how to use it.

How to Use It

1. Scoop out enough of the hand cleaner to be able to coat your hands and forearms.
2. Rub it over the grimy areas, and continue to rub until you see the stains lift.
3. Use a cotton rag and wipe away the cleaner and the grim.

4. Wash your hands under warm water to remove any residual cleaner, and dry them.
5. Don't forget to put the lid back on your hand cleaner so that it doesn't dry out.

Project 42
Make Your Own Green Furniture Polish

When you dust and polish your wooden furniture, you are attempting to accomplish several different things:

1. Remove the dust
2. Remove built-up grime
3. Protect the wood

You want a cleaner, polish, and protectant for your wooden treasures. Therefore, we need to formulate a cleaner that does all three. To do this, we're going to make a cleaning emulsion without soap. We can't use too much water on wood surfaces safely, so if we had soap in our cleaner, it would build up and cause problems. To make our cleaner, we need a solvent such as vinegar and a small amount of water to dissolve, clean, and lift the dirt water away. To protect the wood, we need an oil that can penetrate and not feel greasy to the touch. And we also need an emulsive agent (protein) that can help the oil and solvents work together and protect and clean the wood itself.

Before attempting to make an emulsive agent, it is important to understand how emulsions works. The key to making and emulsion is to avoid having the components of the emulsion separate back into the components. No matter how much you mix oil and vinegar together, it will always separate (break) unless you include something such as an egg yolk as a stabilizer. The lecithin in the egg yolk acts like detergent in dissolving both the oil and the vinegar components. Salad dressings are made this way. Yup, we're going to make a salad dressing for your wood. In fact, many of our parents or grandparents even used mayonnaise as furniture polish in the olden days.

WHAT YOU'LL NEED

- A high-speed mixer with a wire wisp (wisp only)
- A mixing bowl
- Measuring spoons
- A measuring cup
- Cotton rags
- Four jumbo egg yolks (separate the whites and save for cooking)
- 2 teaspoons (9.8 milliliters) of real lemon juice (concentrate only)
- 1 cup (240 milliliters) of soybean oil (vegetable oil) (Figure 8-40)

Figure 8-40

Let's Start

1. Separate the yolks from four jumbo eggs (egg whites destroy the emulsion), and place the yolks in the mixing bowl (Figure 8-41).
2. Add the 2 teaspoons (9.8 milliliters) of real lemon juice. The ratio of lemon juice to oil

Figure 8-41

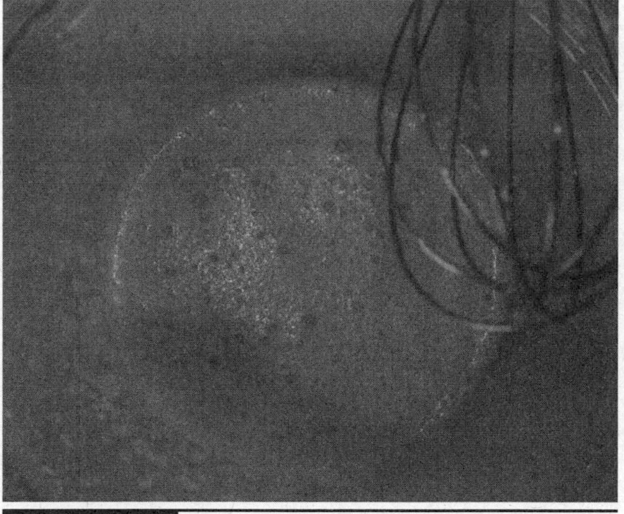
Figure 8-43

has to be exact. Please do not guess (Figure 8-42).

3. Mix the yolks and lemon juice together using the high-speed mixer with a wire wisp until the mixture is frothy (Figure 8-43).
4. Next, start the wisp spinning at maximum speed, and as slow as you can, drip in the oil. This should take several minutes. If you add the oil too fast, the emulsion will be lost. Once all the oil is in, blend a few seconds longer just to make sure that any oil on the sides of the mixing bowl gets included in the emulsion. When you are done, the emulsion should look like creamy frosting. If it comes out runny, start over with a new batch. Remember that the recipe for an emulsion must be followed exactly (Figure 8-44).

Figure 8-42

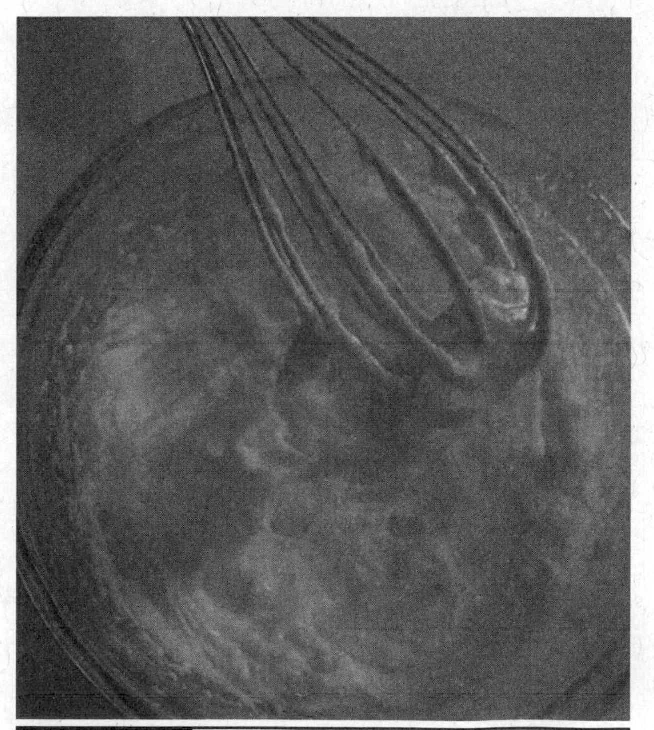
Figure 8-44

5. Store the emulsion in the refrigerator to keep it fresh.
6. Remember to label the container so that you know what it is and how to use it.

How to Use It

1. Dip a clean cotton rag into the cleaner and polish emulsion, and with a small amount on the rag, gently rub it on the wooden surface.
2. Work it around the area until you see or feel the slickness of the wood stop changing.
3. Now take a clean rag and wipe off any residue that didn't work into the wood's surface. Repeat these steps on all your wooden surfaces except your floors because this may make them too slick and unsafe to walk on.
4. Don't worry about the egg and vinegar odor. It won't last long when the air hits it; oxygen neutralizes it in minutes.
5. Don't add any fragrance oil to it because it may destroy the emulsion.

Project 43
How to Remove Stains

I put together a list of stain types I often get in my home and have provided a few ways to remove or treat those stains. I also list the names of conventional cleaners as well as green cleaners you may have made yourself. Even if you have to use a cleaner that may not be natural or green, you still can make a positive difference. When you save something from being thrown away and *add value to it* or *increase its useful life*, you make a difference in the balance of things. Just remember to follow the manufacturers' recommendations, warnings, and directions.

Know Your Stains

The type of stain you have tells you how to remove it. Thus, if you know the stain type, you can fight that stain.

Common Stains

- Protein
- Plant-derived tannin
- Oil
- Pigment
- Combination of pigment and oil
- Combination of protein and pigment
- Unique

What to Do if You Don't Know What the Stain Is

If you don't know what the stain is, see if it has a smell you recognize. The location of the stain is also key, and the color may give you a clue. On the front of a shirt, it's probably a food item; if it is green and on the knees of your jeans, it is probably a grass stain. You get the idea.

Since the appropriate removal method varies with the stain, start by using the least destructive stain removal method first. If the whole item can be submerged, start by soaking the stain. If not, use warm water and spot-treatment technique. Next, use a pH-balanced detergent and lukewarm or hot water, rinse, and let air dry. If you suspect that the stain is iron rust, treat it with rust remover before bleaching. If stain persists, use a pretreatment spray or solvent and an all-fabric nonchlorine bleach. If the all-fabric bleach is ineffective on the stain and the garment is colorfast or white, try a dilute solution of liquid chlorine bleach.

CAUTION **Never mix ammonia and bleach in the same wash load. Toxic fumes are produced.**

Mechanics of Stain Removal

To effectively remove a stain, you need to take a hands-on approach, and knowing what motions to make is just as important as knowing what the stain is made of. If you use the wrong motion, such as rubbing instead of dabbing, you could smear the stain and make it worse. If you use hot wash water, you could "cook in" or "set" the stain, so knowing what to do is important.

Stain-Fighting Verbs

- *Blot*—to dry with an absorbing agent or remove with absorbing material
- *Dab*—to pat or tap gently, such as with something soft or moist, to apply by light strokes
- *Launder*—to wash in water
- *Pretreat*—to treat in advance or as part of a preliminary treatment
- *Rinse*—to wash off soap or remaining dirt or clean with some chemical process
- *Rub*—to move over something with pressure
- Scrape—to remove an unwanted covering or a layer from something
- *Scrub*—to clean with hard rubbing or to wash thoroughly
- *Sponge*—to soak up with a sponge or wipe as to clean or moisten

Know Your Protein Stain Types

- Baby food and spit-up
- Milk and creamer
- Baby formula
- Mucous
- Blood
- Cheese sauce
- Black mud
- Cream
- Pudding
- Egg
- Urine
- Feces
- Vomit
- Gelatin
- White glue
- Ice cream

How to Clean Basic Protein Stains

Newly acquired protein stains can be removed by soaking and agitating in cold water before washing. You can bet that even these stains contain other ingredients besides protein, so they may need a pretreatment first. By all means please refrain from using hot water to soak protein stains; it effectively "cooks" and sets the stain in, causing coagulation between the surfaces of what is stained, thus making the stain more difficult to remove. If protein stains are dried or old, scrape or brush off any crusted matter, and then soak in cold water using a detergent or a pretreatment. After treating the stain, launder as usual in warm (never hot) water, rinse, and inspect. If the stain remains, pretreat again, soak an additional hour, and then rewash (and repeat if necessary).

Fresh Blood Stains

Use cold water or club soda. Sponge the stain immediately, and dry with a soft absorbent terry towel. Repeat as necessary.

Dried Blood Stains

Dab on a little hydrogen peroxide to the dried blood before washing. Test a small section of the material because the hydrogen peroxide also could have a bleaching effect. If the stain persists, rub with some table salt.

1. Mix 1 teaspoon of a pH-balanced detergent with 1 cup of ice-cold water.
2. Blot on the dried blood.
3. Rinse, and if the stain is still visible:
4. Mix 1 tablespoon of household ammonia with ½ cup of ice-cold water.
5. Blot on the stain.
6. Sponge the stain with clean ice-cold water.
7. Launder per label directions.

White Glue and Glue Stick Stains

Try to catch the stain while it's still wet so that it can be removed more easily.

1. Blot the stain with a clean white cloth to remove any excess glue.
2. Mix 1 teaspoon of a mild pH-balanced detergent with 1 cup of lukewarm water.
3. Blot the mix on the glue stain.
4. Rinse, and if the stain still persists:
5. Mix 1 tablespoon of household ammonia with ½ cup of water.
6. Blot the mix on the stain.
7. Sponge with clean water.
8. Launder as usual.

Egg Stains

Try to catch the stain as soon as possible so that it can be removed more easily.

1. Blot the stain with a clean white cloth to remove any excess.
2. Mix 1 tablespoon of household ammonia with ½ cup of water.
3. Blot this on the egg stain.
4. Rinse, and if the stain is stubborn:
5. Mix 1 teaspoon of a mild pH-balanced detergent with 1 cup of lukewarm water.
6. Blot on the stain.
7. Sponge with clean water.
8. Launder as directed.

Dog Feces Stains

Removing a feces and urine stain first by picking and soaking up the waste material as quickly as possible. Using a mixture of 2 tablespoons of ammonia to 1 cup of water, blot the stain to remove any residue, and rinse with cold water. Repeat if all the stain is not removed. To remove odors from dog feces, sprinkle baking soda on the stain. Let the baking soda sit overnight, and then vacuum.

Cat Feces Stains

Removing a feces and urine stain first by picking and soaking up the waste material as quickly as possible. Using straight household white vinegar, blot the stain to remove residue, and rinse with cold water. Repeat if all the stain is not removed. When completely dried, use 1 teaspoon of liquid detergent to 1 cup of water, and thoroughly cleanse the area, blotting it dry. The vinegar smell will dissipate in a few days.

TIP **Pets in our homes use urine to "spray" and mark their territory or have accidents, both of which can cause strong odors. The odor-causing parts of urine are a urea salt and can be neutralized by vinegar to effectively kill that odor.**

Ice Cream Stains

Try to catch the stain while it's fresh so that it can be removed more easily.

1. If the stain is fresh, remove the ice cream, and blot the area as dry as you can.
2. Mix 1 teaspoon of a mild pH-balanced detergent with 1 cup of lukewarm water.

3. Blot this mix on the ice cream stain.
4. Rinse, and if the stain is still visible:
5. Mix 1 tablespoon of household ammonia with ½ cup of water.
6. Blot on the stain.
7. Sponge with clean water.
8. Launder as usual.

Milk Stains

Milk stains can be removed from machine-washable garments by quickly rinsing them with cool water as soon as possible or soaking the garment in cool water for 30 minutes or more.

1. Mix 1 teaspoon of a mild pH-balanced detergent with 1 cup of lukewarm water.
2. Blot the mix on the stain.
3. Rinse and if the stain won't lift:
4. Mix 1 tablespoon of household ammonia with ½ cup of water.
5. Blot onto the stain.
6. Sponge with clean water.
7. Launder as directed on the label.

Know Your Basic Tannin (Plant-Derived) Stains

- Alcoholic beverages
- Beer
- Berries (cranberries, raspberries, strawberries)
- Coffee
- Felt-tip watercolor pen or washable inks
- Fruit juice (apple, grape, orange)
- Tea
- Tomato juice

How to Clean Basic Tannin Stains

These stains are usually removed by detergent washing in warm to hot water during laundering without any pretreatment. If you do pretreat, use only pH-balanced detergents; do not use straight bar soap, soap flakes, or dish detergents on these kinds of stains. The nature of those soaps can and will make these stains permanent or more difficult to remove.

Beer Stains

Beer stains are easy to remove. Remember, though, that dark beers will cause worst stains because of their darker coloring. If the stain is still wet, just soak up as much of the spill as you can with a clean white cloth or paper towel.

1. Mix 1 teaspoon of a mild pH-balanced detergent with 1 cup of lukewarm water.
2. Blot the mix on the beer stain.
3. Rinse, and if the stain is still present:
4. Mix ⅓ cup of white household vinegar with ⅔ cup of water.
5. Blot the mix on the stain.
6. Rinse with clean water.
7. Launder as usual.

TIP **Never rub a stain; always blot it so as not to spread it further.**

TIP **Always rinse stains from behind, when possible, to allow the water to push the stain away.**

Coffee Stains

Coffee stains can be removed easily using this method. Try to catch the stain while it's still wet so that it can be removed easily.

1. Blot the stain with a clean white cloth to remove any excess.

2. Mix 1 teaspoon of a mild pH-balanced detergent with 1 cup of lukewarm water.
3. Blot the mix onto the coffee stain.
4. Rinse, and if the stain persists:
5. Mix ⅓ cup of white household vinegar with ⅔ cup of water.
6. Blot onto the stain.
7. Sponge with clean water.
8. Launder as usual.

NOTE Carpeting also can be cleaned using this method and ingredients.

TIP pH is potential hydrogen; the p stands for potential, and the H stands for hydrogen. A neutral pH is 7, whereas a pH of 1 is acidic (acid) and a pH of 14 is alkaline (base).

TIP Use cloth towels, not paper, to clean; blotting can make paper towels fall apart.

Juice Stains

Remove juice stains by quickly rinsing with cool water. Don't use detergent on most juice stains because it will set the stain and make it impossible to remove. Instead, use white vinegar and blot to remove. Clear or bright-colored fruit juices, such as grapefruit, orange, and apple, can be removed by blotting out as much of the juice as you can and then rinsing with cool water. If needed, use a sponge and white vinegar to clean the spot, and then use a stain pretreater and launder as normal per label instructions. On dry-clean-only items, sponge on a little white vinegar, and rinse with cool water. If the stain persists, it's best handled by a dry-cleaning professional.

1. Blot up the liquid.
2. Rinse with white household vinegar.
3. Blot the the juice stain with vinegar.
4. Rinse with clean water.
5. If stain remains, blot with alcohol.
6. Rinse with clean cool water.
7. Launder as usual.

TIP Use only white vinegar to clean stains.

TIP Use 70% isopropyl alcohol (clear rubbing alcohol) for stain removal.

Soda Pop Stains

Soda pop can be removed by using 3% hydrogen peroxide mixed at the ratio of ¼ cup hydrogen peroxide to 3 cups of water. Using a spray bottle, spray the mixture on, and let it stand for 10 minutes. Rinse with 1 part of white vinegar to 3 parts cold water, and blot dry.

1. Mix 1 teaspoon of a mild pH-balanced detergent with 1 cup of lukewarm water.
2. Blot the mixture onto the soda pop stain.
3. Rinse, and if stain is still visible:
4. Mix ⅓ cup of white household vinegar with ⅔ cup of water.
5. Blot the mixture onto the stain.
6. Sponge with clean water.
7. Launder as usual.

TIP Soaking the stain in near-boiling hot water also will remove pop and juice stains from clothing if the stain has not set in.

Hydrogen peroxide can be found in the health and beauty aids departments as an antiseptic first aid treatment for minor injuries. The oxygen it produces on contact with some organic materials has a non-chlorine-bleaching effect on most stains.

Soy Stains

Coffee stains also can be removed easily using this method. Try to catch the stain while it's still wet so that it can be removed more easily.

1. Blot the stain with a clean white cloth to remove any excess.
2. Mix 1 teaspoon of a mild pH-balanced detergent with 1 cup of lukewarm water.
3. Blot the mixture onto the soy stain.
4. Rinse, and if stain is still present:
5. Mix 1 tablespoon of household ammonia with ½ cup of water.
6. Blot this mixture onto the stain.
7. Rinse again with clean water.
8. Launder as usual.

TIP **When using powdered detergent, use warm water to help it dissolve into your pretreatment or cleaning mix.**

Tea Stains

Tea stains can be removed by dipping a sponge or cotton rag in white vinegar and dabbing it on the stain. Wash the garment as you normally would. Inspect it after washing, and don't throw the garment into the dryer because the heat will set any remaining stains. If the stain persists, rub with salt. Rinse, and launder again as usual.

1. Use white household vinegar.
2. Soak the tea stain in it.
3. Rinse, and if stain is still resistant:
4. Rub with table salt.
5. Launder as usual.

Wine Stains

Wine stains can be easily removed using this method. Try to catch the stain while it's still wet so that it can be removed more easily.

1. Blot the stain with a clean white cloth to remove any excess.
2. Mix 1 teaspoon of a mild pH-balanced detergent with 1 cup of lukewarm water.
3. Blot the mixture onto the wine stain.
4. Rinse, and if the stain is still visible:
5. Mix ⅓ cup of white household vinegar with ⅔ cup of warm water.
6. Blot onto the stain.
7. Launder or sponge with clean water.

Carpeting also can be cleaned using this method and ingredients.

Know Your Basic Oil-Based Stains

- Automotive oil
- Hair/body oil
- Bacon fat
- Hand lotion
- Butter/margarine
- Lard
- Automotive grease
- Mayonnaise
- Collar/cuff greasy rings
- Salad dressing
- Cooking fats and oils
- Suntan oil or lotion
- Face creams and makeup removers
- Diesel fuel

How to Clean Basic Oil Stains

Oil stains can be removed by pretreatment with a heavy-duty liquid detergent or a detergent-based pretreatment. Work the full-strength heavy-duty liquid detergent into the stain or pretreatment product, and then wash the garment using hot water using the recommended amount of detergent for a regular laundry load and rinse. Repeat as many times as needed if removal is incomplete the first time. Remember to inspect for oil stains in bright light so that you can see them easily, especially before drying (machine drying thermosets stains in).

Butter Stains

Remove excess butter as quickly as possible. Make a paste of half laundry powder and half water, and work it into the stain. Launder per care label instructions. Ammonia also will work well to clean grease stains such as butter, but be careful to ensure that it is not going to bleach out the color of the garment. Test on an unseen area of the item for colorfastness.

1. Sponge with a small amount of cleaning solvent or uncolored gel-type hand cleaner.
2. Blot the solvent onto the butter stain.
3. Mix 1 teaspoon of a mild pH-balanced detergent with 1 cup of lukewarm water.
4. Blot this mixture onto the solvent and stain.
5. Sponge with clean water.
6. Launder as you would normally.

TIP **Water sets oil stains. Avoid it before using solvent.**

TIP **Most treatments can work on set-in grease stains as well as fresh ones, but the sooner you tackle the stain, the easier it will be to remove.**

Grease Stains

Sometimes just applying plain old dish soap to a grease stain will do a fine job of removing it. Just work a little soap into the stain, being careful not to spread the grease. Let the garment sit for at least a few minutes, and then launder as normal. Liquid laundry detergent, dishwashing detergent, and even shampoo also can be used in this way. You also can make a thick paste out of powdered laundry detergent and water to pretreat the stain. If you have a whole load of greasy laundry, adding up to 1 cup of ammonia to the wash can help to remove all the stains.

1. Sponge with a small amount of cleaning solvent or uncolored gel-type hand cleaner.
2. Blot the solvent onto the grease stain.
3. Mix 1 teaspoon of a mild pH-balanced detergent with 1 cup of lukewarm water.
4. Blot this onto the solvent and stain.
5. Sponge with clean water.
6. Launder as usual.

Hand Lotion Stains

Use the hottest water setting your washer has when washing out a grease stain. This gives the detergent the boost it needs to remove the grease. Check each time after washing to make sure that the stain is completely removed before drying the garment. A stain that has gone through the dryer will be much more difficult to remove.

1. Sponge with a small amount of cleaning solvent or uncolored gel-type hand cleaner.
2. Blot the lotion stain with solvent.
3. Mix 1 teaspoon of a mild pH-balanced detergent with 1 cup of lukewarm water.
4. Blot this on the solvent and stain.
5. Sponge with clean water.
6. Launder as directed on the label.

Margarine Stains

Absorbing as much oil from the stain as possible helps the stain removal process go more smoothly.

This can be done using an absorbent powder. Cornstarch, baby powder, and even baking soda are products to use for this purpose. Simply apply the powder to the stain, working it in very slightly if necessary. Let it sit for a few hours or even overnight, and then brush it off. This method is especially useful when dealing with more delicate fabrics that may not be able to handle harsher treatments.

1. Sponge with a small amount of cleaning solvent or uncolored gel-type hand cleaner.
2. Blot the solvent onto the margarine stain.
3. Mix 1 teaspoon of a mild pH-balanced detergent with 1 cup of lukewarm water.
4. Blot this onto the solvent and stain.
5. Sponge with clean water.
6. Launder as usual.

Know Your Basic Pigment Stains

- Cherry, blueberry
- Color bleeding in wash (dye transfer)
- Felt-tip pen (permanent ink—may not come out)
- Grass
- Red mud or clay
- India ink
- Powdered drink mixes
- Fresh-cut grass
- Mustard
- Child-safe paints

How to Clean Basic Pigment Stains

Dye stains are very difficult and often can't be removed. First, pretreat the stain, and soak the stained garment in a dilute solution of nonchlorine bleach. If the stain persists, and the garment is white or colorfast, soak in a dilute solution of liquid chlorine bleach and water. Bleaching damage to colored garments is irreversible.

Candy Stains

Candy stains are removed by using a good presoak and laundering in warm soapy water. If the stain persists after a general washing, presoak again, and add a few drops of ammonia. Allow the garment to soak for 30 minutes, rinse well with cool water, and launder per label instructions.

1. Mix 1 teaspoon of a mild pH-balanced detergent with 1 cup of lukewarm water.
2. Blot the mixture onto the candy stain.
3. Rinse, and if stain is still present:
4. Mix ⅓ cup of white household vinegar with ⅔ cup of water.
5. Blot onto the stain.
6. Sponge with clean water.
7. Launder as directed.

Fruit Stains

1. Blot up all liquid as dry as possible.
2. Mix 1 teaspoon of a mild pH-balanced detergent with 1 cup of lukewarm water.
3. Blot the mixture onto the fruit stain.
4. Rinse with clear white household vinegar.
5. Blot with the vinegar.
6. Sponge with clean water, and if stain remains:
7. Blot with isopropyl alcohol.
8. Launder as directed on the label.

Iodine Stains

1. Mix 1 teaspoon of a mild pH-balanced detergent with 1 cup of lukewarm water.
2. Blot the mixture onto the iodine stain.

3. Mix 1 tablespoon of household ammonia with ½ cup of water.
4. Blot this on the last attempt to remove the stain.
5. Mix ⅓ cup of white household vinegar with ⅔ cup of water.
6. Blot again on the stain.
7. Sponge with clean water.
8. Launder as directed on the label.

Marking Ink Stains

1. Sponge with a small amount of cleaning solvent or gel-type hand cleaner.
2. Blot onto the ink stain.
3. Mix 1 teaspoon of a mild pH-balanced detergent with 1 cup of lukewarm water.
4. Blot this onto the solvent and ink stain.
5. Dab with rubbing (isopropyl) alcohol.
6. Sponge with clean water.
7. Launder as directed on the label.

Since bleaches can alter the color of a fabric as well as the stain, bleach the whole garment, and do not try to bleach the stained spot alone.

Know Your Combination Pigment and Oil Stains

- Ballpoint ink
- Candle wax
- Carbon paper
- Carbon typewriter ribbon
- Crayon
- Eye makeup (mascara, pencil, liner, shadow)
- Floor wax
- Furniture polish
- Lipstick
- Livestock paint
- Pine resin
- Shoe polish
- Tar

Crayon Stains

Crayons resist water and water-soluble cleaners. Warm and hot water soften and tend to spread crayon stains, making the stain worse.

Method 1

1. Carefully scrape up or lift off as much solid crayon as possible. Be careful not to spread the stain during removal.
2. Place a tray full of ice cubes into a zip-type freezer bag and set it over the stained area for several minutes. The ice makes the crayon more brittle. Again scrape off residue.

Method 2

1. Use a hair dryer to soften the crayon. Once again, carefully scrape up as much solid crayon as possible.
2. An alternative to a hair dryer is a clothes iron. Place a clean absorbent material, such as a few layers of paper towels or a napkin, over the stain. Apply a warm, not hot, iron over the layer of absorbent covering. Melt and liquefy the crayon so that it will be absorbed into the covering material.
3. Treat the stain with a small amount of cleaning solvent or gel-type hand cleaner. Allow it to stand for about 30 minutes. Agitate with a scrub brush. Rinse thoroughly with water. Once satisfied with results, allow the surface to completely dry.

Cheese Stains

1. Mix 1 teaspoon of a mild pH-balanced detergent with 1 cup of lukewarm water.
2. Blot the mixture onto the cheese stain.
3. Rinse, and if the stain is still visible:
4. Mix 1 tablespoon of household ammonia with ½ cup of water.
5. Blot this on the stain.
6. Sponge with clean water.
7. Launder as specified on the label.

Mascara Stains

1. Sponge with a small amount of cleaning solvent or uncolored gel-type hand cleaner.
2. Blot the solvent onto mascara stain.
3. Mix 1 teaspoon of a mild pH-balanced detergent with 1 cup of lukewarm water.
4. Blot this on the solvent and stain.
5. Sponge with clean water.
6. Launder as usual.

Paint Stains

1. Sponge with a small amount of cleaning solvent or uncolored gel-type hand cleaner.
2. Blot the solvent onto the paint stain.
3. Mix 1 teaspoon of a mild pH-balanced detergent with 1 cup of lukewarm water.
4. Blot this onto the solvent and paint stain.
5. Sponge with clean water, and if the stain persists: Seek a professional carpet cleaner/dry cleaner.

Paste Wax Stains

1. Sponge with a small amount of cleaning solvent or uncolored gel-type hand cleaner.
2. Blot the solvent onto the wax stain.
3. Mix 1 teaspoon of a mild pH-balanced detergent with 1 cup of lukewarm water.
4. Blot the mixture onto the solvent and stain.
5. Sponge with clean water.
6. Launder as directed on the label.

Tar Stains

1. Sponge with a small amount of cleaning solvent or uncolored gel-type hand cleaner.
2. Blot the solvent onto the tar stain.
3. Mix 1 teaspoon of a mild pH-balanced detergent with 1 cup of lukewarm water.
4. Blot this mixture on the solvent and stain.
5. Sponge with clean water.
6. Launder as directed.

Ink Stains

Ballpoint inks, especially blue, consist of so many various "ink recipes" that it is impossible to know the various solvents, resins, and oils in just one. Most inks are removable through experimenting with different solvents. Some inks are indelible. Your situation may be hopeless, leaving permanent or lightened stains.

1. Treat a well-folded absorbent white cloth or towel with a uncolored gel-type hand cleaner. Dab the affected area with the cloth, and blot with a dry towel simultaneously.
2. Treat a well-folded absorbent white cloth or towel with isopropyl rubbing alcohol. Dab the affected area with the cloth, and blot with a dry towel simultaneously.
3. Treat a well-folded absorbent white cloth or towel with nail polish remover or acetone. Dab the affected area with the cloth, and blot with a dry towel simultaneously
4. Treat a well folded-absorbent white cloth or towel with turpentine. Dab the affected area

with the cloth, and blot with a dry towel simultaneously

5. If one solvent cleaner does not solve the problem, move down to the next in the order these cleaners are given.

TIP Never apply solvents directly onto the garment. Use solvents only by first applying them to the item with which you are dabbing the stain.

Lipstick Stain

Lipstick is a dye in an oily base, so a solvent-based remover is the only type that will remove it.

1. Rub vegetable or mineral oil onto the affected areas, and allow it to soak for 15 minutes.
2. Blot excess oil with an absorbent paper towel or terrycloth rag.
3. Sponge the remaining affected areas liberally with isopropyl rubbing alcohol or uncolored gel-type hand cleaner.
4. Mix 1 teaspoon of a mild pH-balanced detergent with 1 cup of lukewarm water.
5. Blot this mixture onto the solvent and lipstick stain.
6. Launder as specified on the label.

TIP Water-based or wet-spot treatments will only spread and set lipstick stains.

Know Your Combination Protein and Pigment Stains

- Barbecue sauce
- Ketchup or tomato sauce
- Cocoa or chocolate
- Face makeup (powder, rouge, foundation)
- Gravy

Chocolate Stains

1. Mix 1 teaspoon of a mild pH-balanced detergent with 1 cup of lukewarm water.
2. Blot this mixture onto the chocolate stain.
3. Rinse, and if the stain still persists:
4. Mix 1 tablespoon of household ammonia with ½ cup of water.
5. Blot this mixture onto the stain.
6. Repeat step 1.
7. Sponge with clean water.
8. Launder as directed.

Gravy Stains

1. Sponge with a small amount of cleaning solvent or uncolored gel-type hand cleaner.
2. Blot with the solvent.
3. Mix 1 teaspoon of a mild pH-balanced detergent with 1 cup of lukewarm water.
4. Blot this mixture on the solvent and the stain.
5. Sponge with clear water.
6. Launder as usual.

Ketchup and Mustard Stains

Condiment stains, such as ketchup and mustard, can be removed by first blotting up the stain as quickly as possible. Then sponge the stain with cool water, work a little liquid detergent into the stain gently with your fingers, and rinse in cool water. Apply a good stain remover, such as Shout, to the stained area, and launder per care label instructions. Never dry a garment in the dryer until you are sure that the stain is fully removed. This will set the stain permanently, and it will never be removed completely.

1. Mix 1 teaspoon of a mild pH-balanced detergent with 1 cup of lukewarm water.
2. Blot this mixture onto the stain.

3. Mix 1 tablespoon of household ammonia with ½ cup of water.
4. Blot onto the first stain treatment.
5. Sponge with clean water.
6. Launder as directed on the label.

Know Your Unique Stains

Chewing Gum

Apply ice to harden the gum. Crack or scrape off the excess. Spray with a pretreatment aerosol product. Rub the material together aggressively with a heavy-duty liquid detergent. Rinse with hot water. Repeat if necessary. Launder.

Deodorant

Apply liquid detergent, and wash in warm water. Built-up or caked-on aluminum or zinc, which are often the active ingredients in deodorants, may be impossible to remove.

TIP **Soak caked-on deodorant stains in straight clear household vinegar overnight.**

Fingernail Polish

For organic fabrics, use nail polish remover with acetone on the stain, and then do a spot treatment.

TIP **Do not use nail polish remover (or acetone) on synthetic fabrics because they will dissolve. Take these fabrics to a professional dry-cleaner, and identify the stain to the proprietor.**

Mold/Mildew

Mold and mildew are living, growing organisms. They basically eat organic fibers, causing damage, weakening, and total destruction of fibers and fabrics. Shake or brush the item outdoors, and be careful to wear appropriate breathing protection (an N-95 respirator mask) so as to not breathe the potentially toxic and allergen-prone mold or mildew. Pretreat dark stains with heavy-duty liquid detergent. Launder in hot water with a heavy-duty detergent.

TIP **If you are doing serious cleanup in a moldy environment, wear full protective gear with gloves, safety glasses, a Tyvek suit over your clothing, and boots or boot covers on your feet. Be sure to remove and dispose of the Tyvek suit and footwear as soon as you leave the moldy area.**

TIP **Never take contaminated protective gear back into a car or living quarters.**

Paints

For latex paints, treat while still wet. Soak in cold water, and wash in cool water with a heavy-duty detergent. After paint has dried 6 to 8 hours, removal is very difficult.

Smoke and Ash

Shake off excess ash outdoors. Launder using a heavy-duty detergent (as recommended by the manufacturer), 1 cup of water conditioner, and ½ cup of all-fabric bleach. Use a water temperature appropriate for the fabric. Inspect and repeat if necessary.

Urine

Rinse in cold water and launder. For stains, sponge with a cloth using a detergent solution, rinse with a cloth using a vinegar solution, and let air dry; if an odor remains, sprinkle with baking soda, and wait 24 hours before vacuuming off the baking soda.

How to Clean Odds and Ends

Cleaning Aluminum

Aluminum is a metal that's lightweight and often has a bright silvery appearance when it's not coated, anodized, or painted. Other metals are often added to aluminum to make it harder or less porous to fluids such as oils. Aluminum beverage cans are lined with resin so that they don't leak. Aluminum reacts with oxygen to make a resistant oxide that fights corrosion and attack by most mild chemicals.

To brighten dingy aluminum utensils, try using a mild acid such as vinegar. Mix 2 tablespoons of vinegar with each quart (946 milliliters) of water in a pan. Add the utensils, and boil for 10 minutes or so. Remove hard water and mineral deposits from tea kettles by boiling in equal parts of white vinegar and water for several minutes and letting them stand an hour or so (repeat if needed). Remember to rinse with plain water before using the tea kettle.

Burned-on Food and Grease

Fill your sink with hot soapy water, and let the item soak for 1 hour. Scrape off as much food as possible with a dull item such as a wooden spoon, plastic spatula, or plastic scrubber. Finish with a soap-filled steel wool pad.

TIP **Use caution with abrasive cleaners (e.g., scouring powders, steel wool, abrasive polishes, etc.) because they may permanently scratch aluminum; painted or anodized aluminum surfaces will be permanently damaged.**

TIP **Prevent warping and cracking by not cleaning aluminum when it is too hot to touch or when temperatures are near freezing.**

Outdoor Surfaces

First, remove bugs and tree sap as soon as possible. Suitable solvents will remove tar and similar substances. Test the solvent first if the aluminum is painted to be sure that it doesn't also remove the paint. Clean off dirt and dust with soap and water.

Cleaning Brass

Brass is an alloy of copper and zinc. It tends to oxidize or "tarnish" easily when exposed to the oxygen in the air. Brass is often given a clear coating of lacquer to prevent oxidation. When dealing with "raw," uncoated brass instead of finished, lacquered brass, the reaction between the raw metal and chemicals can create even greater oxidation conditions.

Light Soils

Use 70% isopropyl (rubbing alcohol) applied with a soft sponge, and rub lightly.

Heavier Soils

Dampen the side of a sponge with water, and apply a light-scouring, low-abrasion cleaner. Work the product into the sponge, and rub it lightly on the item. Once complete, wipe the surface thoroughly with a clean cotton rag.

Polishing

Wipe down brass with a cloth dampened with olive oil, and then buff it dry with another soft cotton rag. The trace amount of oil in the cloth should not smear or discolor the brass, especially after buffing. Olive oil retards tarnish.

Polishing Antique Brass

To polish antique brass pieces, wash in hot, soapy water to remove grime and wax; then rinse and dry. Moisten a soft cotton cloth with boiled linseed oil, and rub it on the brass surface until all the dirt and grease have been removed. Now polish the brass with a clean soft cotton rag.

Polishing Antique "Old" Brass

Very old brass items, especially if they are in poor condition, require special care. For a soft finish, wash in hot soapy water, rinse, and dry. Apply boiled linseed oil with a soft cloth, and rub to remove tarnish. Wipe off excess polish with a clean cloth.

Heavy Tarnish, Stains, and Corrosion

Wash in hot soapy water or a weak ammonia and water solution and rinse. Dampen a soft cloth in hot vinegar, and then dip in table salt and rub the brass, or make a paste of salt and vinegar. You may need several applications. When the item is clean, wash in hot soapy water, rinse, and dry thoroughly; then polish with a slice of lemon dipped into table salt, and rub over the corroded area. Wash, rinse, and dry carefully.

Cleaning Bronze

Bronze today is basically an alloy of copper and any metal except zinc. It is generally more expensive than brass and more corrosion-resistant. Bronze forms a green color that is protective of the metal and is often seen on artwork. Reproduced, it is called verde bronze. Bronze will deteriorate rapidly if it is exposed to moisture and corrosives such as chlorides or sulfides.

Solid bronze often is lacquered to protect the finish. Lacquered bronze only needs dusting and an occasional wiping with a damp cloth.

Keep bronze pieces as clean as possible. Accumulations of dust and dirt can eat into the metal surface. Dust regularly using a soft cloth. Do not rub too vigorously, especially on any protruding parts. If a bronze piece has been neglected for a long time and is covered with grime, thoroughly clean it with a soft brush. Remove all dust from crevices, and then lightly rub the entire surface with a soft cotton rag.

Cleaning Cast Iron Cookware

Cast iron items for food preparation have been used longer in our homes than nonstick and aluminum cookware. Cast iron is heavy, so it distributes heat slowly and evenly; it's great for frying fish or chicken.

Although cast iron cookware is normally virtually indestructible, it requires special care and cleaning to maintain its properties and remain rust-free. With proper care, cleaning, and storage, cast iron cookware can provide lifetimes of use. Take good care of your cast iron cookware, and it can be passed down and enjoyed by your children's children.

Seasoning Cast Iron

New cast iron cookware requires seasoning before first use, and it's typically silver or gray in appearance. The seasoning process is basically the use of heat and vegetable or animal oil to create a organic natural polymer that makes a smooth protective moisture-proof finish. Unlike modern nonstick skillets, the seasoned finish is repairable. Seasoning is an easy process, and with proper care, your cast iron cookware should never require seasoning again. Some cast iron pieces are labeled as seasoned, and the durable factory-seasoned finish doesn't require additional seasoning unless it has been scratched or damaged by rust and moisture.

How to Season Cast Iron

1. Wash the skillet with hot soapy water. Dry thoroughly.
2. Rub a liberal coat of lightweight clean cooking oil over the entire skillet, inside and out, including the handle.

3. Place the skillet in an oven set on low temperature (about 250 to 300°F). Position the skillet upside down in the oven with aluminum foil under it to catch any oil drippings.
4. Wear oven mitts so that you don't get burned, and remove the skillet after about 10 to 15 minutes; let it cool. The oil should be tacky to the touch over the entire inside surface. If you have wet spots, smooth them out. If you have spots that are not covered with a tacky surface, add oil to those spots. Place the skillet back in the oven until you have the entire inner surface covered with a tacky oil coating. If you have trouble with this step, increase the oven temperature and allow more time.
5. Once this is accomplished, turn up the oven setting to 500°F. Allow the item to cure for 1 hour in the oven before you check it.
6. When cured properly, your skillet will have an even, black, shiny sheen over the entire inner surface.
7. If you have a shiny sheen, you are finished. If not, put it back in the oven, raise the temperature to 550°F, and continue cooking as long as it takes to get the sheen. Since oven temperatures vary, the time required for this step may not be the same for all ovens.

TIP Your kitchen will be smoky from the oil, so do this on a day when you can ventilate the kitchen.

TIP Avoid using your cast iron cookware to prepare tomato-based foods or other acidic foods containing lemon juice or vinegar. They will deteriorate the nonstick surface, and as a result, the cookware might require additional seasoning.

TIP Don't store leftovers in cast iron. Food and moisture will deteriorate the seasoned surface and cause it to rust.

Cleaning Seasoned Cast Iron

While the cookware is still warm, place it in an empty sink under hot running water. Wipe the surface with a dishcloth or soap-free kitchen sponge. If there are stuck-on or burned-on foods, use coarse salt as an abrasive to aid in removal. Rinse the clean cookware, and dry it thoroughly. Traces of moisture will cause rusting. To ensure that the cookware is completely dry, set it on the stove or in a warm oven for a few minutes to dry any excess moisture.

Cleaning Copper

Copper is valued for strength, malleability, ductility, and the ability to conduct electricity and heat.

Decorative items should be kept clean and dusted. Copper is sensitive to air and oxidizes or "tarnishes" faster in moist air. Most pieces of modern decorative copper are protected by factory-baked lacquer. Only dusting and an occasional washing with lukewarm soapy water are needed to keep lacquered objects shiny.

Copper Pots

To remove tarnish from copper pots, rub with lemon halves dipped in salt. You also can make a paste with salt and lemon juice and apply it with a cotton rag and rub.

Copper Utensils

Copper interiors never should be used for acidic foods because toxic compounds can form if food is cooked, stored, or served in copper containers. Even if copper pans are lined with tin, they should not be used for acidic foods such as fruits, fruit juices, salad dressings, tomatoes, and vinegar-containing foods. Copper bowls may be used for beating egg whites or copper kettles for cooking high-sugar foods such as fudge because these foods are alkaline. Utensils with copper on the bottom or outside and stainless steel, aluminum, or porcelain enamel on the interior are safe to use and conduct heat well. Avoid high heat, which discolors copper bottoms.

Ideally, you should clean copper bottoms after each use, even though the tarnish does not affect cooking results or the pan's efficiency. Do not use an abrasive cleaner or steel wool to clean copper bottoms.

Cleaning Copper Utensils

Wash tarnished copper utensils with soap and warm water, and polish with a cleaner of equal parts of salt and vinegar. After rubbing the item, wash it, rinse thoroughly, and dry it. If copper is tarnished, boil the article in 1 quart (946 milliliters) of water with 1 tablespoon of salt and 1 cup of white vinegar for several hours. Wash with soap in hot water. Rinse and dry. Make a paste of lemon juice and salt, and rub with a soft cloth, rinse with water, and dry.

Cleaning Marble

Marble is stone that is generally polished and used in fine building work, furniture, or decorative art. It may be white or colored. It is porous and easily stained. Marble is easily etched by acids. Wipe off anything spilled on marble immediately, as you would from a wood surface.

Marble Furniture

Marble may be stone, but it is porous and stains easily. Wipe off anything spilled on marble immediately, just as you would from a wood surface. Use coasters under beverage glasses to avoid moisture rings.

Regular Cleaning

Occasionally wash marble surfaces with lukewarm water, and wipe dry with a clean cloth. Wipe the surface with a damp cotton rag, and try not to leave streaks. Once or twice a year, depending on the amount of soil, hand wash with a mild detergent and warm water, rinse, and wipe dry. A light coat of wax will protect the surface of marble but is not considered essential. Use a colorless wax. Don't wax white marble because it may yellow. A marble sealer can be applied to clean marble and will protect it from staining and allow soil to be wiped off with a damp cloth.

Cleaning Porcelain Enamel

Porcelain enamel is a specially formulated, highly durable glass permanently fused to metal under extremely high temperatures. The metal may be steel, cast iron, or aluminum. It may be a protective surface for cooking utensils, kitchen and bathroom fixtures, and appliances such as dishwashers, ranges, refrigerators, washers, and dryers. Range tops and ovens are often made of porcelain enamel because there is no good substitute presently available to withstand the high temperatures. Older appliances may have very thick porcelain enamel surfaces.

Appliances

Wash with detergent and warm water, and rinse. Do not use abrasive pads or scouring powders because these will scratch the glassy surface.

Decorated Enamelware

Wash in sudsy water, and dry with a soft cloth.

Bathroom Fixtures

These can be cleaned with detergent and hot water or with a foam bathroom cleaner. Avoid using household cleaners that contain abrasives. Porcelain enamel bathtubs made since 1964 are acid-resistant. Older ones may be dulled by acid contact or other chemicals spilled in the tub.

Kitchenware

Wash in sudsy water, and use a plastic scouring pad or wooden scraper to remove burned-on food. Burned-on food may be loosened by soaking in a solution of 2 teaspoons of baking soda and 1 quart (946 milliliters) of water.

Sinks

Sinks can be cleaned with detergent and hot water or with a foam bathroom cleaner. Avoid using household cleaners that contain abrasives. Do not leave acidic foods sitting on the sink surface for a long time. Much of the porcelain used on cast iron sinks made before 1964 was not acid-resistant. But many acidic fruits such as lemons and other citrus fruits, cranberries, and other acidic foods such as vinegar and salad dressing possibly could etch porcelain enamel if left in the sink for a long time.

Cleaning Silver

Sterling silver is an alloy of 92½ percent silver and 7½ percent copper. Plated silver is silver that has been electroplated over another metal. Silver tarnishes when exposed to air. Store silver items in treated paper or cloth or plastic film.

Silver has enemies. Rubber severely affects silver. Rubber corrodes silver, and it can become so deeply etched that only a silversmith can repair the damage. Raised designs can be lost permanently. Avoid using storage cabinets or chests with rubber seals, rubber floor coverings, rubber bands, etc.

Other enemies of silver include table salt, olives, salad dressing, eggs, vinegar, and fruit juices. Serve these foods in china or glass containers.

How to Remove Tarnish

Apply a paste of baking soda and water. Rub, rinse, and polish dry with a soft cloth. To remove tarnish from silverware, sprinkle baking soda on a damp cloth, and rub it on the silverware until the tarnish is gone. Rinse and dry well.

Place a sheet of aluminum foil in the bottom of a pan, add 2 to 3 inches of water, 1 teaspoon of baking soda, and 1 teaspoon of salt, and bring to a boil. Add the silver pieces, and boil for 2 to 3 minutes, making sure that the water covers the silver pieces. Remove the silver, rinse, dry, and buff with a soft cloth. This method cleans the design and crevices of silver pieces.

Coat the silver with toothpaste, then run it under warm water, work it into a foam, and rinse it off. For stubborn stains, use an old soft-bristled toothbrush to scrub them out.

Silverware Care

Silver is easily scratched, so never use harsh abrasives. It is corroded or tarnished by salt and salt air, sulfur, and rubber. Frequent use deters the formation of tarnish. Do not let silver stand with food on it; salty or acidic foods can stain it. Rinse it if it will not be washed at once. Store silverware in a chest lined with flannel or in a airtight plastic bag when not in use.

Cleaning Silverware

Wash in warm sudsy water. Rinse well, and dry immediately.

Washing in a Dishwasher

Silverware may be washed in a dishwasher. Use the automatic rinse dispenser feature of the dishwasher. The rinse agent lowers the surface tension of the rinse water so that it sheets off the surfaces, and droplets don't form. Remove new silverware from the dishwasher immediately after the last rinse cycle, and towel dry.

Cleaning Stainless Steel

Stainless steel is an alloy of nickle, iron, and copper that contains more than 10 percent chromium. Stainless steel resists stains but occasionally dulls or will show oily fingerprints. This steel is noted for its hardness and is used for utensils, tableware, sinks, countertops, and small appliances. In the process of making it, a little of the chromium in the alloy is used to form a hard oxide coating on the surface. If this is taken off, through corrosion or wear, the steel rusts like regular steel.

Cleaning Stainless Steel

1. Rub stainless steel sinks with olive oil to remove streaks.
2. To clean and polish stainless steel, simply moisten a cloth with undiluted white or cider vinegar and wipe clean. This also can be used to remove heat stains on stainless steel cutlery.
3. Remove streaks or heat stains from stainless steel by rubbing with club soda.

Stainless Steel Flatware

Wash by hand or in the dishwasher. Rinse off foods if the stainless flatware is not to be washed soon.

Utensils

Do not let pans boil dry or overheat on the burner; this causes discoloration. Stainless steel pans on burners do not distribute heat evenly, and foods tend to stick in "hot spots," so careful stirring of foods is important. Pans with a copper bottom or a layer of aluminum or copper hidden in the bottom help to overcome this problem by distributing the heat better.

Wash by hand or in a dishwasher. If washed by hand, rinse well and polish dry at once with a soft dish towel to avoid spots and streaks. Do not use harsh abrasives or steel wool on stainless steel. Cooked-on food or grease can be removed from stainless steel utensils with a fine abrasive cleaning powder or a paste of baking soda and water.

Sinks

Wash with a solution of dishwashing liquid and water or a solution of baking soda and water. Rinse and polish dry with a paper towel or cotton rag. You can brighten a sink by polishing with a cloth dipped in vinegar, or sprinkle a little baking soda on a sponge and rub the sink gently, rinse, and dry.

CHAPTER 9

Home-Friendly Green Pest and Weed Control Projects

FOR MANY HOME OWNERS who deal with pests, a good offense is better than a good defense. If you have bugs, find out what they are. The same is true about plants or weeds. There are many books and guides to help you identify your weeds and insects, or find someone at your local nursery to help you. The first thing to remember is to treat only the problem areas rather than trying to spray and treat your entire home or garden. Select a control measure that targets the pest. Use the weed's or insect's own biological makeup against itself. Living things cannot become resistant to their own biology. It's like a fish trying to breathe air or me trying to breathe water; biology is key to controlling weeds, insects, and other pests.

Control Issues

The key to safely controlling pests, whether plant, insect, or animal, is to use a minimal amount of effort or substance to control the pest without affecting the biological order of your own environment. You want to discourage, repel, stop the reproduction of, or destroy those pests and not affect the beneficial insects, plants, or animals by doing so. Most insects are either beneficial or cause no harm. Some, like bees and butterflies, are vital to the fertilization process for plants. Most ant species collect plant seeds and insect eggs. Ants are like "nature's little vacuum cleaners." Blackbirds prey on slugs and snails, dragonflies eat mosquitoes and aphids, and even the common ground beetle feeds on cutworms. Ladybugs, spiders, lacewings, and garter snakes are also natural pest control agents and are creepy crawlies that definitely should be welcomed around your home.

Pesty Pesky Weeds

It is clear that some weeds are plants that present a major threat to commercial interests or are superweeds that are called an *invasive species* and can invade and often take over and destroy natural habitats. Invasive species are usually introduced plants with few natural enemies in the area of invasion. By forming dense colonies, they displace native plants and reduce biodiversity. The more ordinary kinds of weeds are not considered a significant threat to the natural environment because they tend to disappear after artificial disturbances have been removed, whereas invasive species are dreaded visitors. Invasive species are introduced through many means. Intentional introductions often have been for agricultural or ornamental purposes. Once introduced, some of these species escape their enclosures or cultivation areas and can become established as viable populations. Accidental introductions are usually the result of contaminated freight or movement of contaminated wood products (including shipping pallets, bracing, and cardboard), plants, or food products.

Weed the People

The plants we sometimes call weeds are entitled to a bit more respect than they have received from many of us. Even in these modern times of the organic and natural gardening boom, we all have had a tendency to underestimate the value that the native plants that we call weeds have for the environment. Some native weed species are like daring pioneers and take root in degraded areas where the soil is damaged or was nearly destroyed by natural disasters or humankind's actions. Pioneer weed species are necessary to the healing process of the landscape because their decaying organic matter improves the quality of the soil and sets the stage for the success of other, preconceived, more valuable plants. These pioneering plants also can quickly cover exposed soil that is at risk of erosion caused by wind and water. The quick growth of "weed" plants can prevent the erosion of valuable soil from occurring and in time can be replaced by other slower-growing plants. Weeds are also the same native plants that are sources of food and cover to various kinds of mammals, birds, and insects. For example, Monarch butterfly caterpillars feed on the common milkweed, which makes the insect taste bad to birds. Thus the milkweed is a source of both food and defense. Every weed has the potential to be an important source of fiber, energy, food, and even medicine yet undiscovered. Native weeds species often make their way into our yards and gardens because that's what they do, and those native weeds were here long before you were even born. Therefore, the next time you pull, pluck, pick, stomp, chop, or spray a weed dead, don't feel proud or guilty about what you just did. You may have won a battle, but those native weeds and all their families had already won the war long before you became the invasive species to them and wiped out their cousin.

Project 44
Make Your Own Weed Killer

Weeds can take a lot of extra work to pull, pluck, and chop away. Having a chemical means to stop them cold is a viable option for most of us, especially if the chemical is safe and easy to use. More often than not, a homemade weed killer can be more affordable than those store-bought types. In fact, you may have all the ingredients for this weed killer in your home already and don't even know it.

Weed Killer Active Ingredient 1

Vinegar (acetic acid) has long been used in cooking and cleaning and for a host of other applications, and its potential use as a herbicide is underrated or little known. Vinegar can be produced naturally by decomposing plant products with microbes under anaerobic conditions. Household gourmet vinegar is usually made from wine (grapes), cider (apples), or grains (starches). The sugars in these plant products are first converted into ethanol (alcohol) and carbon dioxide through microbial fermentation. In the second process, another group of bacteria (acetobacter) converts the alcohol portion to acid. This is the acetic, or acid, fermentation that forms vinegar. Petroleum-based ethanol (hydrolysis of ethylene) also can be used to make vinegar, so buyer beware if you are all about natural or organic. Check out your vinegar's source.

Regular household vinegar is around a 5 percent acetic acid concentration, whereas pickling vinegar is 10 percent acetic acid. While this strength of acid works on some weeds, a greater concentration may be needed for other, more mature weeds (so we need to add another active ingredient). Plants (weeds) are sensitive to acids because they affect the activity of all the enzymes in the plant and the availability of plants from which they get nutrients and quickly die. However, vinegar's effects on treated weeds in surrounding environments are less severe than other weed killers because vinegar easily dilutes and degrades in water and soil. Therefore, the negative effects of vinegar don't bioaccumulate, and they decrease the pH of the soil for only a few hours. Vinegar (dilute) is a safe, natural, and biodegradable product by all means.

Weed Killer Active Ingredient 2

Alcohol (ethanol or isopropyl) is a poison for most living things; the potency of the poison varies based on the type of alcohol and how much of it is introduced. When people drink ethanol, the intoxication effect, or buzz, caused by it is the result of the deliberate poisoning of their body. Fortunately, a healthy person's body can filter out the poison after a few hours, and all that remains after the self-induced event is a hangover. Because of the effects of alcohol, some people get addicted to it and can slowly poison themselves to death, or the inexperienced alcohol abuser can drink too much the first time and cause themselves sudden death. Other alcohols are even more toxic. Methanol, for instance, if ingested, can produce the buzz or intoxication effect as well, but as the body deals with the poisoning, the liver makes an even greater toxin, formaldehyde (used to embalm dead bodies). This pickling effect causes blindness and eventually death for those who consume methanol.

Plants and weeds also can be poisoned by alcohol. It easily disrupts their cell activity and also robs their metabolism of the moisture they need to make energy. Alcohols evaporate quickly, and ethanol in particular biodegrades very easily when exposed to air. We're going to use isopropyl alcohol (rubbing alcohol) because it's inexpensive, easy to find (health and beauty aid sections of stores), and bonds with water better than other alcohols do.

Weed Killer Inactive Ingredient 3

Soap (surfactant) is able to break down the protective waxy surface found on most weeds. This makes the plant more susceptible to the action of the weed killer's active ingredients. Soap also lowers the surface tension of water, so the mix can stick to the surfaces of weeds rather than running off. This allows more of the killing agent to hang on and get working.

Weed Killer Inactive Ingredient 4

Raw egg is a protein that works well as an adhesive to further smother a plant. Both the yolk and the white have bonding properties and have been used as adhesives throughout history. Ancient Egyptians used egg whites to adhere gold leaf, and it is believed that native Americans used eggs as an adhesive in building canoes. Egg white, known as albumin, is waterproof, once dried, and in warm summer air, it takes very little time to dry. If you've ever had to remove a dried egg stain, I'm sure you understand how good a hold it can have.

WHAT YOU'LL NEED

- Vinegar (5 to 10 percent concentration)
- Liquid dish soap (base cleaner)
- Isopropyl alcohol
- A raw egg
- A 2-cup (500-milliliter) measuring cup
- A hand-pumped sprayer
- A tablespoon (Figure 9-1)

Figure 9-1

Let's Start

1. With the measuring cup, add 4 cups (1,000 milliliters) of vinegar to the sprayer.
2. Now add 1 cup (250 milliliters) of isopropyl alcohol to the sprayer.
3. Now add 2 tablespoons (30 milliliters) of the dish soap (base cleaner).
4. Finally, crack open the raw egg, and add it to the sprayer (compost the shell).
5. Close the sprayer, and give it some real good shakes to break up the egg yolk and mix everything as best you can. Follow the manufacturer's directions for using your spray pump. You're ready to go.

CAUTION **Just because this is a natural weed killer doesn't mean that it can't be harmful if misused. This weed killer is nonselective, and this natural weed killer can harm your grass and those prized plants as well. Don't spray on windy days; the wind can carry your spray where you don't want it to go.**

How to Use It

Never pull the trigger to spray until you're right up close to the targeted weed. You'll want to saturate the weed's leaves and exposed surface areas during good dry, sunny weather. The real damage to the sprayed weeds takes place about 2 days after application, when the sun hits the leaves. If it rains, you'll need to reapply the spray because the rain can wash away your weed-killing efforts. For larger and more mature weeds, expect to reapply several times to get results because weeds with established root systems are able to help the exposed and damaged parts of the weed heal quickly. Nevertheless, if you keep attacking the top side of the weed, the roots run out of energy eventually, and the weed dies completely.

Making Your Own Creepy Crawly Controls

Many people are aware that controlling pests according to organic principles means what we're using is free from synthetic or potentially harmful pesticides, but that is only one small part of natural pest control. We also want to protect the health of the soil and the ecosystem in which we live and where our children are raised. Natural pest control is a hands-on and a know-how type of labor of love with the reward of having some peace of mind and pride in what you do, and it is often less expensive than conventional methods of control. Saving green money is a good shade of green indeed.

Remember, It's a Bug Eat Bug World Out There

Let's say that you sprayed something to kill mosquitoes around your property and in so doing you unknowingly killed off your local dragonfly population. You may have just made your mosquito problem worse by mistakenly killing off the dragonflies that ate their own weight in mosquitoes every night. Your best bet would've been to repel or draw the mosquitoes away from your home and letting the dragonflies feast on them.

NOTE It's also important to keep the mosquito population under control by removing their breeding grounds:

- Remove standing water in rain gutters, old tires, buckets, plastic covers, toys, or any other container where mosquitoes can breed.
- Change the water in birdbaths, fountains, wading pools, rain barrels, and potted-plant trays at least once a week.
- Drain and fill puddles of water with dirt.

Too broad of a spectrum of insect control measures may cause a larger infestation by wiping out the beneficial insects that feed on the target insect, controlling its population naturally. A more natural way to tip the balance would've been to reduce the pest's own natural habitat and then increase or enhance the beneficial insect's habitat.

Making Your Own Pest Controls

Most modern pesticides are based on formulas using once-natural ingredients that were passed down through the generations because the chemistry was right. Going back to the original, naturally derived ingredients is a way to make pesticides that work that do not pollute and may save you money as well. Mixing and matching environmentally-friendly green pesticides, you can easily and simply transform your personal environment into a healthier space. Less toxic pest control can prove to be gratifying by ensuring that your family's health is protected and your home is a place to rest and find peace. As an added bonus, ounce for ounce, making your own formulas can cut your cost.

The Many Uses of Boric Acid

Boric acid is a naturally occurring compound that contains three elements: boron, hydrogen, and oxygen. It's represented chemically as H_3BO_3. This colorless, odorless powder consists mostly of borates and is found naturally in soil, rocks, and seawater. It is mild acid that is used for medicinal, industrial, and pesticide purposes. Boric acid (boron) is naturally present in many foods, such as fruits, vegetables, and grains. Thus an average person consumes some amount of boron every day in his or her diet. The borates used to make boric acid are derived from dried salt lake beds found in deserts and other arid regions such as the Mojave Desert in California.

Medicinal

Boric acid is usually used as an antiseptic for treating minor burns and cuts because it inhibits the growth of microorganisms in the body. Boric acid eyewash is considered to be an effective remedy for pink eye (conjunctivitis) and other eye infections. It is also used in contact lens solutions and other eye disinfectants. The antibacterial, antifungal properties of boric acid allow it to be used in acne treatments and for the prevention of athlete's foot and even diaper rash.

Industrial

Boric acid's primary industrial use is in the preparation of Pyrex glass and fiberglass. It has its uses in the jewelry industry to prevent some side reactions caused by oxidation that heated metals might undergo. Even in manufacturing fireproof clothes, boric acid is used commonly as a fire retardant.

On farms, many fertilizers contain boric acid powder as an active ingredient. Boric acid powder can be used in fireworks to make a green flame. Boric acid is also used to treat wood because it's known to prevent rot caused from fungi, microorganisms, and insect attack.

Pest Control

Boric acid is one of the oldest inorganic compounds known to humankind as a treatment for pests. It is nature's proven long-term treatment in eliminating cockroaches, palmetto bugs, water bugs, ants, silverfish, carpenter ants, and termites. Boric acid generally is known as a desiccant, so basically it kills by removing the moisture from the body of the pests that consume it, causing severe dehydration, which will affect electrolyte balance. Boric acid is a stomach poison, and it also can enter the blood by inhalation. Boric acid is an acid that can decrease the pH level with the possible effects of renal, respiratory, and cardiovascular failure. Boric acid can be toxic to humans and our pets; it's simply a matter of quantity (like pure water is safe until it's above your nose).

Boric acid is no more poisonous than table salt or aspirin, which are basically poisons that we use every day that in small amounts add to our value of life. Life would be bland without salt, and humankind's headaches and strokes wouldn't be as easy to treat without aspirin. The fact that such substances may be poisonous is moot compared with the value they provide. To be poisoned by salt or aspirin, an actual effort would need to be involved, so the risk is low. However, it still deserves to be respected. Boric acid is important and is promising for use as a pesticide and fungicide for the control of fungi, termites, roaches, and other insects, as well as being a wood preservative. With appropriate amounts and application methods, boric acid can be used as a safe alternative to a long-lasting pesticide or fungicide without any negative effects on the environment.

Shh ... It's a Secret

Boric acid is the "secret ingredient" in many store-bought treatments for insect control because it's inexpensive and deadly to all insects. It can attack an insect's nervous system, as well as being a drying agent to their bodies. Boric acid is often used for controlling cockroaches in homes, restaurants, and public buildings. It works in small amounts and retains its potency for long periods, provided that it remains dry. Boric acid also has no repellent effect on insects, so roaches often return to treated areas repeatedly until they die. Pure (99 percent) boric acid powder is odorless and nonstaining. Boric acid can be obtained from general and hardware stores as powdered roach killer. Look at the active ingredients for 99 percent boric acid. You also can buy it at a pharmaceutical grade from pharmacies and drugstores, as well as in bulk from any taxidermy supplier (boric acid is used as an animal skin preservative).

How to Use Boric Acid

Success with boric acid is all about application. For best results on hard surfaces, the powder should be applied in a very thin layer that's barely visible to the naked eye. Piles will be avoided by crawling insects on the move. You want them to walk through the boric acid so that they pick it up on their bodies so that it can be ingested as the insect preens and cleans the powder from its legs and antennas. Some of the boric acid is also absorbed through the covering of the insect's body. All species of cockroaches are susceptible to boric acid, provided that the powder is applied to areas where the roaches are found.

NOTE *Inhalation*—Causes irritation to the mucous membranes of the respiratory tract. Boric acid may be absorbed from the mucous membranes and, depending on the amount of exposure, could result in the development of nausea, vomiting, diarrhea, drowsiness, rash, headache, fall in body temperature, low blood pressure, renal injury, coma, and death.

Ingestion—Symptoms parallel those of absorption via inhalation. Adult fatal dose is reported to be 5 to 30 grams or more.

Skin contact—Causes skin irritation. Boric acid is not absorbed significantly through intact skin. It is readily absorbed through damaged or burned skin. Symptoms of skin absorption parallel those of inhalation and ingestion.

Eye contact—Causes irritation, redness, and pain.

Chronic exposure—Prolonged absorption causes weight loss, vomiting, diarrhea, skin rash, convulsions, and anemia. The liver and particularly the kidneys may be susceptible. Studies of dogs and rats have shown that infertility and damage to testes can result from acute or chronic ingestion of boric acid. Evidence of toxic effects on the human reproductive system is inadequate.

Aggravation of preexisting conditions—Persons with preexisting skin disorders or eye problems or impaired liver, kidney, or respiratory function may be more susceptible to the effects of boric acid.

TIP Always read the manufacturer's warnings, and follow the directions on the label.

Inhalation—Remove to fresh air. If the person is not breathing, give artificial respiration. If breathing is difficult, give oxygen. Call a physician.

Ingestion—Induce vomiting immediately as directed by medical personnel. Never give anything by mouth to an unconscious person.

Skin contact—Remove any contaminated clothing. Wash the skin with soap or mild detergent and water for at least 15 minutes. Get medical attention if irritation develops or persists. Wash clothing before reuse.

Eye contact—Immediately flush the eyes with plenty of water for at least 15 minutes, lifting lower and upper eyelids occasionally. Get medical attention immediately.

Project 45
Make a Puffer to Kill Roaches

To apply boric acid powder accurately in a nearly invisible layer, we need to build a puffer applicator. Some boric acid products are sold in a squeeze bottle that can be used as a puffer, but you may find this project a little more user-friendly, and the puffer can be used get behind heavy appliances and hard-to-reach spots.

WHAT YOU'LL NEED

- An empty squeezable ketchup bottle
- Boric acid powder
- A drinking straw
- Duct tape

Let's Start

1. Recycle a clean and dry ketchup bottle, and remove the lid with the flip top open or removed (remove the flip cap by holding it and twisting the bottle).
2. Push a drinking straw into the small opening in the lid. You may have to crimp it a little to get it to start through the hole in the lid.
3. Now tape the straw to the lid as best as you can (be careful not to get the tape inside the lid, making it impossible to screw it back on later).
4. Now fill the bottle about a third of the way full with boric acid powder.
5. Put the lid back on, and the puffer is ready to use.

6. Apply a fine layer by shaking the container and giving it a quick small squeeze to puff out the fine boric acid powder into the area where insects crawl. Give the bottle a shake each time after you puff it, and refill your puffer only a third of the way up each time so that the powder can move.
7. Remember to label the container so that you know what it is and how to use it.

How to Use It

Puff it around baseboards and under and behind the refrigerator, stove, sink, dishwasher, washer, and dryer. Puff it into openings around drain pipes and electrical conduits, in cracks and crevices, along baseboards, and in corners of cabinets, cupboards, and closets. The insects that walk through the dust will ingest it and die within hours.

Die Roach Die!

Cockroaches prefer to live in cracks, crevices, and secluded areas close to food, moisture, and warmth. Kitchens and bathrooms are the most common areas to find cockroaches, although any area of a home may become infested, especially if the infestation is severe or if species other than the German cockroach are involved. Key areas for treatment include under/behind the refrigerator, stove, and dishwasher; in the openings where plumbing pipes enter walls (such as under sinks and behind the toilet, shower, and washing machine); and to cracks along edges and corners inside cabinets and pantries. A place where roaches like to hide is behind plugged-in adapters used by small electric appliances. These electrical devices are often warm and attract young roaches that even live inside the electric outlet inside the wall. Don't forget the hollow space under kitchen and bathroom cabinets, which can become a breeding ground and hiding place for cockroaches. This area can be accessed and treated by injecting powder through any opening at the top of the mop board (kick plate); if there is no opening, try drilling a few small holes close to the top where no one can see them.

Project 46
Make a Boric Acid Flea Control

Fleas are tiny wingless insects that suck the blood of animals, such as dogs, cats, rats, chickens, and humans. They live in a host fur or feathers and feed from their hosts' skin and blood. Flea bites are very itchy and painful to both humans and animals. Many people are allergic to flea saliva, which causes a rash on their skin. These tiny wingless parasites also live in carpets because they mistake the carpet fiber for fur. They can't survive long in carpets without a food source, so they need to latch onto new living hosts. These insects like to live in dark, moist areas such as underneath furniture or behind doors. Check both sides of your carpets for fleas that cling to the fabric, and also inspect large pieces of furniture such as couches because fleas like to live in things we humans live on.

WHAT YOU'LL NEED
■ A recycled plastic container with a lid
■ A vacuum cleaner
■ A sharp knife
■ Boric acid powder
■ Baking soda (Figure 9-2)

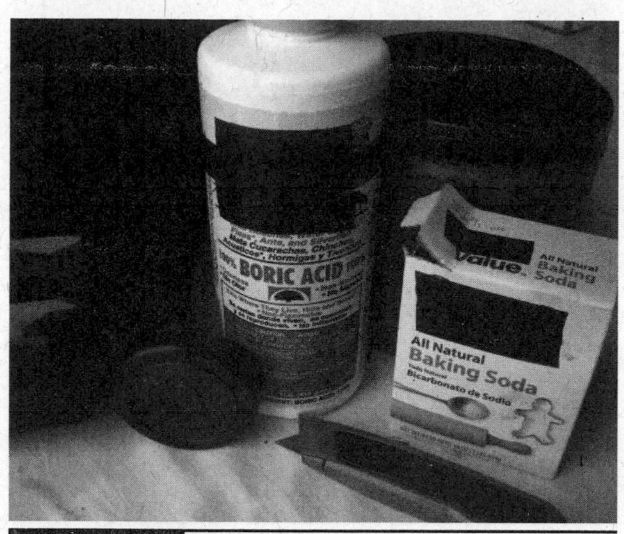

Figure 9-2

Let's Start

1. Mix equal parts baking soda and boric acid in a plastic dish, filling it about half full.
2. Put the lid on the plastic container, and carefully using the knife cut slits or small holes in the lid so that you can sprinkle the powder through them.

How to Use It

1. First, clear all areas that require treatment by removing all movable items from the areas to be treated, such as shoes, toys, and so on.
2. Now vacuum all the carpeted surfaces thoroughly to remove dirt and dust. This will enable the boric acid powder to act more effectively.
3. All cushions also should be removed from furniture and vacuumed.
4. Clean and vacuum the walls as well.
5. Now sprinkle the boric acid mixture lightly over all carpeted areas, including closet floors and under furniture. Pay special attention to favorite resting places of your pets because these are likely to be areas where fleas are abundant.
6. Using a broom, brush the boric acid mixture into the carpet fibers to spread the powder evenly until no visible powder remains on the carpets.
7. Normal vacuuming can be resumed 24 to 48 hours after application. All vacuum bags should be removed and discarded immediately after vacuuming. The boric acid remains active for up to a year.
8. Remember to label the container so that you know what it is and how to use it.

TIP **Boric acid also kills flea larvae but is not as effective at killing the adults, so you may not see the results for 2 to 6 weeks while the adult population dies off. As such, it is helpful to vacuum frequently to kill the adult fleas during the initial weeks after application.**

TIP **To make use of boric acid powder to kill fleas on furniture, simply sprinkle straight boric acid powder very lightly over the furniture. Using a hand brush, work the powder deep into the upholstery until the powder disappears. Then vacuum off all excess powder from the furniture.**

Project 47
Make a Boric Acid Ant Control

Like human beings, ants are social creatures, living together in a cooperative manner. Every ant colony consists of a variety of different-sized worker ants that do the day-to-day labor and one or more queens. Ants are important components of the ecosystem because their soil-moving skills make them important in nutrient cycling, and they are a food source for many animals. Ants can be grouped by their method of getting food. The feeding guilds include seed harvesters, liquid feeders, predators, slave-making ants, and omnivores (the ants that eat almost anything).

Most ants nest underground, but during the reproductive season, an ant colony will produce winged males and females that usually swarm above ground to mate.

Ants are a common household pest during the warm and wet months, when they're on the prowl for food and hives are expanding. To kill most ants in your home, it's as simple as using a mixture of boric acid and sugar water. The sugar in the mixture attracts the ants; the boric acid slowly poisons them after they eat it. There are many different types of ants, so you may need to try increasing or decreasing the amount of boric acid in the mixture to find the right potency for your particular ants. The proportion of boric acid to sugar water is key. If you make the mixture too weak, it won't kill anything, but if you make it too strong, it will kill the foragers before they can get back to the queen and deliver the poisoned food to her. To end the war with ants, the queen must die.

WHAT YOU'LL NEED

- A recycled plastic container with a lid
- A thin piece of cardboard (cereal box type)
- A 2-cup (250-milliliter) measuring cup
- A tablespoon
- Boric acid powder
- Sugar
- Water (Figure 9-3)

Let's Start

1. Add 1 cup (125 milliliters) of water, 2 cups (500 milliliters) of sugar, and 2 tablespoons (30 milliliters) of boric acid to the recycled plastic container.

Figure 9-3

2. Tear some 2-inch (50-millimeter) × 2-inch (50-millimeter) pieces of single-layer cardboard.
3. Put little drops of the mixture on the pieces of cardboard, and place them wherever you've seen the ants. You need to make sure that the ants will find the cardboard, and you must let them eat their fill. The foragers need time to deliver your weapon to the queen. Give it at least 3 days to work, and please resist killing the forager ants (Trojan horses) as they feed, no matter how big of a feeding mass they become.
4. Remember to label the container so that you know what it is and how to use it.

TIP **While boric acid is far less toxic than most commercial chemical pesticides, you still do not want kids or pets ingesting this sugary but tainted mixture.**

Project 48
Making Boric Acid Chalk to Control Crawling Pests

As the title says, we're going to make a chalk line using boric acid that stops or even kills crawling insects if they cross it. I call it "chalk block" because you basically are using it to draw a line that you don't want pests to cross. If the surface the insects are on is a solid rough texture like a brick wall or cement floor, you're good to go.

WHAT YOU'LL NEED

- Three recycled toilet paper roll tubes
- A 2-cup (250-milliliter) measuring cup
- A recycled plastic container
- A tablespoon
- Duct tape
- Cayenne pepper powder
- Boric acid powder
- Plaster of paris
- Water (Figure 9-4)

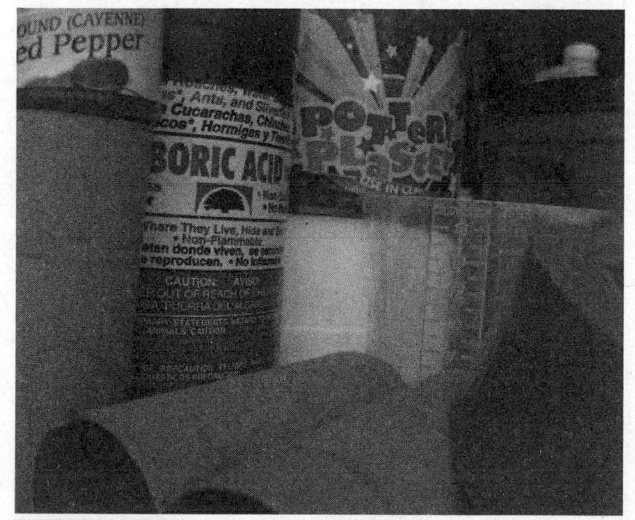

Figure 9-4

Let's Start

1. Add 1 cup (125 milliliters) of dry plaster of paris powder to the recycled plastic container.
2. Now add 2 tablespoons (30 milliliters) of boric acid powder to the container.
3. Follow this with 1 heaping tablespoon (15 milliliters) of cayenne pepper powder (Figure 9-5).
4. Put a piece of duct tape over one of the open ends of each of the three empty toilet paper roll tubes (these are the chalk stick molds) (Figure 9-6).

Figure 9-5

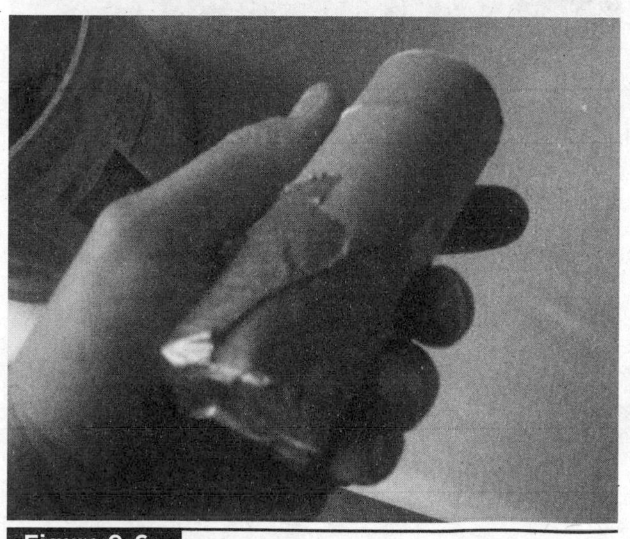

Figure 9-6

5.. Now add water slowly to the contents of the plastic container, mixing it with the tablespoon as you pour. Once you start to get a paste that sticks to the spoon and is lump-free, you're done adding water.
6. Using the tablespoon, fill the toilet paper roll tubes with the paste until they're full. Set them on the taped-over ends and let them cure.
7. Peel the paper away from the toilet paper roll to expose the chalk as you use it to draw your pest-control lines (Figure 9-7).
8. Remember to label the chalks so that you know what they are and how to use them.

TIP These sticks are not toys, so don't let kids play with them.

Figure 9-7

Garlic as a Pest Control

Garlic is a friendly plant to use in the kitchen and can add value to your garden in a variety of ways. The main benefit of garlic for the home gardener is its natural fungicide and pesticide properties; pests naturally stay away from it. If you plant garlic near your prized plants, it may help to keep pests away. You also can use garlic extracts to protect your crops. Garlic contains upwards of 200 known chemical compounds, roughly split among over 100 sulfur compounds (some water-soluble, others oil-based, packed with proteins and carbohydrates).

Project 49
Make a Garlic Concentrate for Multiple Pest Control Use

We're going to make a concentrate from garlic that can be used in insecticide soap, horticultural oil, and a pepper bug spray. This is a concentrate, so a very small amount can go a long way. Moreover, the concentrate is full of water-soluble as well as the organic solvent–derived compounds extracted from the garlic. Therefore, unlike garlic concentrates on the market today, your concentrate is going to carry more of the helpful pest-controlling natural chemical compounds that garlic has to offer. We're using castor bean oil and ethanol (alcohol) to extract the fat- and oil-based compounds from the garlic by exploiting the polar bonds that the chemicals have for one another.

WHAT YOU'LL NEED

- Castor bean oil (from a drug or health food store)
- 190 proof grain alcohol (Everclear)
- Garlic bulbs
- A small saucepan with a lid
- Two large-mouth glass jars
- A hot plate or stovetop
- Coffee filters
- A garlic press or hammer
- A tablespoon
- A funnel (Figure 9-8)

Figure 9-8

Let's Start

TIP Extinguish any open flame or ignition sources when using 190 proof grain alcohol because it is extremely flammable.

1. Add 5 tablespoons (75 milliliters) of castor bean oil to the saucepan.
2. Use the garlic press to smash a whole bulb (or two) of garlic into the saucepan of castor oil. Use the tablespoon to stir the pressed garlic into the castor oil as often as you can to coat the garlic with the oil. You want to restrict the amount of oxygen in the air from prematurely breaking down the garlic's chemical bonds. If the oil gets a little thin, add more—the amount of oil measured isn't as important as keeping the garlic covered with the castor oil.
3. Add a little heat, and slowly stir the garlic and castor oil in the saucepan until it begins to simmer. Remove it from the heat, cover it, and let it stand for 24 hours.
4. After the 24 hours, scoop the entire contents of the saucepan into the large-mouth glass jar.
5. While looking at the amount of garlic and oil in the jar, add approximately the same volume of 190 proof grain alcohol (if you had around 2 inches of garlic and oil, add around another 2 inches of alcohol). Put the lid on, and carefully give the jar a good shake.
6. Open the second jar, and with the coffee filter in the funnel, set the funnel in the jar's opening.
7. Open the jar that has the garlic, castor oil, and ethanol mixture in it, and slowly pour it into the coffee filter–lined funnel. If the coffee filter is too slow, you can use a clean cotton rag or cheesecloth instead.
8. When as much liquid as possible is filtered into the second jar, return the solid contents back into the first jar, and put the lid on the second jar that has the liquid in it.
9. Again looking at the amount of solids inside the first jar, add approximately the same volume of 190 proof grain alcohol (if you had around 2 inches of solids from the filter, add around another 2 inches of alcohol). Put the lid on, and carefully give it a good shake. Let this sit for 24 hours.
10. After the 24 hours, open the second jar again, and with the coffee filter in the funnel, set the funnel in the jar's opening and slowly pour the contents from the other jar into the coffee filter–lined funnel.
11. When as much liquid as possible is filtered into the second jar, return the solid contents back into the first jar to use somewhere else in your yard.
12. The second jar contains the concentrate. Shake it well before each use.
13. Remember to label the jar so that you know what it is and how to use it.

TIP Use the leftover garlic solids to ward off rabbits and other animals by placing them around prized plants.

Project 50
Make a Garlic Insecticide Soap

Making your own garlic insecticide soap is a great way to save money and keep your vegetable garden under control at the same time. This project works best on soft-bodied pests such as aphids, white flies, and spider mites. These are among the most common garden pests. Insecticidal soaps kill insects by entering the pest's respiratory system and breaking down internal cell membranes. It is effective only when it is wet, so practice your aim well.

TIP Some plants, such as ferns, are sensitive to soap, so don't use a soap mixture on them. New growth on plants may be too tender for soap, so apply sparingly at first. Plants under stress also may have a bad reaction to garlic soap insecticide. Plants that are under stress from drought should be soaked with water the day before you treat them. You always should test your mixture first on just one leaf on a plant. If it is fine the next day, your solution is ready to use. It is better not to spray your plants in the middle of the day.

WHAT YOU'LL NEED
■ Natural liquid soap or the pH-balanced soap from the base cleaner project in Chapter 8
■ Garlic concentrate
■ Water
■ A clean and empty recycled large spray bottle
■ A 2-cup (500-milliliter) measuring cup
■ A tablespoon

Let's Start

1. Add 2 cups (500 milliliters) of water to the spray bottle.
2. Now add 2 tablespoons (30 milliliters) of the garlic concentrate to the water in the bottle.
3. Finally, add 1 tablespoon (15 milliliters) of the liquid soap.
4. Close the spray bottle, give it fa ew good shakes, prime the sprayer by giving it a few pumps, and it's ready to use.
5. Spray it directly onto the insects on your plants.
6. Remember to label the container so that you know what it is and how to use it.

TIP Wet garlic soap insecticide may harm the insects you want to hang around, such as ladybugs, so don't spray it directly on them. After it is dry, it will not harm your beneficial insects.

TIP Many garden pests like to hide underneath the leaves of plants. For best results, aim upward, and get under that foliage. Aim directly at those bugs. You may need to spray your organic pesticide again in a few days if you have a heavy infestation of pests.

TIP For heavy insect infestations, it is best to spray your plants again every few days.

Project 51
Make a Garlic Insecticide Flour Spray

If the garlic fails to deter an insect pest, the flour used in this spray turns into a sticky substance called *dextrin* that can slow down a pest like a sticky trap. Or the insect will ingest it while it cleans itself of the brew, and flour will affect their digestive tracts. Apply it only on sunny days and early in the morning. You always should test your mixture first on just one leaf on a plant. If it is fine the next day, your solution is ready to use.

WHAT YOU'LL NEED
■ Natural liquid soap or the pH-balanced soap from the base cleaner project in Chapter 8
■ Garlic concentrate
■ Finely ground wheat or potato flour
■ Warm water
■ A clean and empty recycled large spray bottle
■ A 2-cup (500-milliliter) measuring cup
■ A tablespoon

Let's Start

1. Add 2 cups (500 milliliters) of warm water to the spray bottle.
2. Now add 2 tablespoons (30 milliliters) of the garlic concentrate to the water in the bottle.
3. Now add 1 tablespoon (15 milliliters) of the liquid soap to the mix.
4. Finally, add 3 tablespoons (45 milliliters) of flour to the bottle.
5. Close the bottle, and give it some good shakes (to dissolve the flour so that it can pass through the sprayer). You'll want to prime the sprayer by giving it a few pumps, and it's ready to use.
6. Remember to label the container so that you know what it is and how to use it
7. Spray it directly onto the insects on your plants.

Project 52
Make a Garlic and Vegetable-Based Horticultural Oil

For over a century, various oils have been used to control unwanted insects, fungi, and mites on dormant (wintering) fruit trees and woody plants. Our project oil, however, can be applied to more types of plants and plants in full leaf throughout the growing season because it's a light oil and it biodegrades easily. Horticultural oils work by blocking an insect's breathing passages, causing it to suffocate. Oil also interferes with the cell metabolism of soft-bodied insects such as aphids, scale, mealybug, and some caterpillars. Oils also can disrupt the feeding patterns of certain disease-carrying insects and even can kill overwintering insect eggs and fungal spores on fruit and ornamental trees.

WHAT YOU'LL NEED
■ Natural liquid soap or the pH-balanced soap from the base cleaner project in Chapter 8
■ Garlic concentrate
■ Vegetable oil (cottonseed, soybean, or sunflower)
■ Water
■ A recycled clean and empty large spray bottle
■ A 2-cup (500-milliliter) measuring cup
■ A tablespoon

1. Add 2 cups (500 milliliters) of water to the spray bottle.
2. Now add 1 tablespoon (15 milliliters) of the garlic concentrate to the water in the bottle.
3. Now add 1 tablespoon (15 milliliters) of the liquid soap to the mix.
4. Finally, add 1 cup (250 milliliters) of vegetable oil to the bottle.
5. Close the spray bottle, and give it some good shakes. You'll want to prime the sprayer by giving it a few pumps, and it's ready to use.
6. Spray it directly onto the insects on your plants.
7. Remember to label the container so that you know what it is and how to use it.

TIP

- **Do not apply on sensitive plants, and avoid drift onto them.**
- **Do not apply on drought-stressed plants. Plants that are under stress may be damaged.**
- **Do not apply during freezing weather or when the humidity is above 90 percent for longer than 36 hours.**
- **Do not apply when plant foliage is wet or when rain is expected.**

Project 53
Make a Pepper Spray to Control Chewing Insects

Hot pepper spray works on bugs, just like it does on bad guys, because in large amounts it can cause great pain and discomfort. The chemical that makes pepper spray a powerful deterrent and a sought-after spice is called *capsaicin*. Capsaicin puts the burn in pepper spray, besides being used in repellent sprays and to protect gardens from insect and animal pests. Chemists are now developing an environmentally safe marine coating made with capsaicin that can stop barnacles, which waste energy and cause damage, from growing on vessels.

WHAT YOU'LL NEED

- Natural liquid soap or the pH-balanced soap from the base cleaner project in Chapter 8
- Finely ground cayenne pepper
- Garlic concentrate
- Water
- A recycled clean and empty large spray bottle
- A 2-cup (500-milliliter) measuring cup
- A small saucepan with a lid
- A large-mouth glass jar
- A hot plate or stovetop
- A coffee filter
- A tablespoon
- A funnel

Let's Start

1. Bring 1 cup (250 milliliters) of water to a roiling boil in the small saucepan.
2. Add ½ cup (125 milliliters) of ground cayenne pepper, and stir it into the boiling water.
3. Remove the pan from the heat, and cover it with a lid.
4. After the pan cools, use the funnel and coffee filter to filter the hot pepper water into the open spray bottle.
5. Now add 1 tablespoon (15 milliliters) of the garlic concentrate you made in Project 49 to the water in the bottle.
6. Add 1 tablespoon (15 milliliters) of the liquid soap to the mix.

7. Finally, add enough water to fill the rest of the bottle.
8. Close the spray bottle, and give it some good shakes. You'll want to prime the sprayer by giving it a few pumps, and it's ready to use.
9. Remember to label the container so that you know what it is and how to use it.
10. Spray it directly onto the insects on your plants.

Lay the pepper powder that was left over from this project out to dry by spreading it out on a flat surface, and use it in your garden. Mammals such as rabbits, deer, and even cats and dogs don't like pepper powder, so wherever you don't want them, sprinkle the pepper powder. Birds aren't bothered by hot pepper powder, so you can add it to their bird feeders if the squirrels won't leave them alone.

Project 54
Make a Castor Oil–Based Mole Control

Moles cruise from lawn to lawn and garden to garden looking for worms and grubs to eat. Moles are carnivores, not vegetarians, so when they're in our lawns and gardens, they want to eat grubs and worms, not our plants or roots. When we have rich, healthy soil, we have healthy plants, and this helps the worms and grubs to prosper, attracting hungry moles. The solution to this problem is to definitely not kill off the worms or to stop improving the soil. Moles are also more active and prefer soils that are damp, so overwatered lawns are the primary target for moisture-loving moles. To reduce mole populations, reduce watering to no more than an inch or less once a week. You also could make and use a mole repellent similar to commercial mole repellents found in garden centers and stores. The common ingredient in professional mole repellents is castor bean oil, which does not kill moles but does encourage them not to stay where not wanted. If you've ever tasted castor oil or understand its laxative effect, you might get the idea why moles don't like castor oil.

WHAT YOU'LL NEED

- Natural liquid soap or the pH-balanced soap from the base cleaner project in Chapter 8
- Castor bean oil (sold at drug and health food stores)
- Garlic concentrate
- Water
- A recycled clean and empty large spray bottle
- A large-mouth glass jar
- A garden sprayer
- A tablespoon

Let's Start

1. In the large-mouth glass jar, mix two parts castor oil to one part liquid soap to one part garlic concentrate. Depending on your mole problem, you may want to make as big of a batch as you can. Just remember the formula: 50 percent castor oil, 25 percent liquid soap, and 25 percent garlic concentrate.
2. Put the lid on the jar, and give it a good shake until it is a frothy white mixture.
3. Remember to label the container so that you know what it is and how to use it.

TIP **You always should test your mixture first on just one part of your yard or garden. If it is fine the next day, your solution is ready to use.**

How to Use It

1. Add 2 tablespoons (30 milliliters) of the mixture to 1 gallon (3.8 liters) of water in your garden sprayer.
2. This treats about 1,000 feet (300 meters) of yard or garden per gallon. Remember to spray in and around mole holes.
3. Don't repeat again for 24 hours.

Project 55

How to Make a Beer Trap for Slugs

Slugs are the one garden pest that personally drive me the most insane. I remember as a kid walking barefoot and stepping on them and how gross it was. Over the years, I've battled those little slime balls, trying to keep them away from my tomatoes. Slugs have their place as the "garbage guts" of nature, and they're food for other creatures such as ground beetles, turtles, toads, frogs, lizards, salamanders, lightning bug larvae, turtles, and garter snakes. Chickens (particularly laying hens) love slugs; they can eat all the slugs they can find. Other slug hunters include blackbirds, crows, ducks, jays, owls, robins, seagulls, and starlings. Slugs are pure protein and important to the ecosystem.

Slugs are hermaphrodites: They all have male and female reproductive systems. They can stretch to 20 times their normal length, enabling them to squeeze through tiny openings to get at food. Slugs can follow slime trails they left from the night before. Other slugs also can pick up on this same trail, creating a slug network to food sources. Slugs, like snails, actually have shells, but their shells are much smaller and not visible because they are underneath the flesh on their back.

Slug eggs are found often in the soil but can pop up just about everywhere. Slug eggs can sit unhatched for years and then hatch when moisture conditions are right. To identify the eggs, look for oval-shaped white-colored eggs in moist soil areas and under rocks and boards. Eggs are laid in clusters of two dozen eggs each. The adults also overwinter in the soil and can live for many years.

TIP **If you step on a slug and the slime is difficult to get off your foot, hands, or anything else, try using white vinegar, and then wash with warm soapy water.**

WHAT YOU'LL NEED
■ Beer, wine, or grape juice
■ Recycled shallow plastic containers
■ A shovel (Figure 9-9)

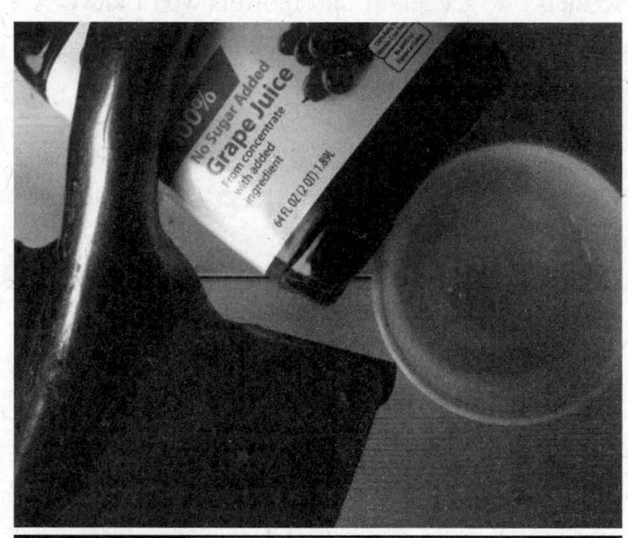

Figure 9-9

Let's Start

1. Find the places where you think slugs are active, and dig a few holes in the soil that are wider and deeper than the containers you're going to use. Put the container in the soil, and place a small stick over the edge of the container so that legged insects can escape the trap if they accidentally fall into it. Some insects are beneficial to the garden.

2. Fill the containers with beer, wine, or grape juice (use the least expensive) until about half an inch from the rim.
3. Wait for the slugs to crawl into the containers overnight. As the beer, wine, or grape juice attracts them, the slugs are drawn into the containers and are unable to crawl out; they will drown.
4. Just empty the containers into your compost pile, dead slugs and all. Refill the trap with fresh bait, and then put it back in the hole.

Stopping Slugs: Rough 'em Up

- Try lava rock as a barrier in areas where plants need protection. I have heard from many people who say that it works very well.
- Try cat litter or oil-dry clay (diatomaceous earth); it works as a barrier but must be replenished after rainfall or each watering.
- Try a mulch made of stems and leaves of strong-smelling herbs such as wormwood, mint, tansy, and lemon balm, along with conifer twigs mixed.
- Try a barrier of human hair, pet fur, or horsehair to entangle the slugs. Using hair also supplies some nitrogen to the soil.
- Try cultivating the soil. Roughing up the soil will kill sleeping slugs and their eggs.
- Try hoeing weeds and clods of soil under which slugs like to hide, and you may expose their eggs.
- Alcohol spray will dry them up. Mix any 100 proof (50 percent or greater) alcohol with 1 quart (1 liter) of water, and spray it on them.

Project 56
Capture and Kill Fleas with Christmas Tree Lights

Flea traps are a good way to rid your home of fleas if safe chemicals such as boric acid are just not an option. Fleas are attracted to heat and motion; needless to say, the animals that fleas feed on are warm-blooded and move. This trap works on those two principles—heat and motion—so we're going to use a light that produces a small amount of heat, and to simulate motion, the light needs to blink slowly. Fleas see the difference in light and dark. In so doing, they assume that it is something walking by, blocking the light, so they jump toward that blinking light. To capture and kill the fleas, we want them to drown in a pan of soapy water. The soap in the water lowers the surface tension of the water so that the fleas can't float on the surface, so they sink and drown.

TIP **Electricity and water can be dangerous together, so keep them apart.**

WHAT YOU'LL NEED

- Liquid soap
- Water
- A cardboard box that's much taller and wider than your water tray or pan
- A string of 100 small blinking Christmas tree lights (non-LED type)
- A tray or pan
- Screwdriver
- Box cutter (Figure 9-10)

Figure 9-10

Let's Start

1. Start by putting holes in the bottom of the cardboard box. Try to keep the holes a couple fingers' width apart from each other. If you have 100 Christmas lights, make 100 holes.
2. Cut openings on three sides of the box four fingers' width from the corners and halfway up the box and across it. You are making doors for the fleas to come into the trap (Figure 9-11).
3. Once you have all the holes made, start putting the lights into the holes (light first), until the bulb is in the box. If you made your holes the proper size, the plastic socket around the bulb should fit snugly in or go through the holes (Figure 9-12).
4. Fill the pan halfway full of water and a few drops of liquid soap.
5. Place the tray or pan of soapy water underneath the box of lights. Plug the lights in, watch the lights blink, and wait (Figure 9-13).
6. Leave the bowl under the light overnight. In the morning you will want to empty the water trap of dead fleas and make a fresh

Figure 9-12

Figure 9-11

Figure 9-13

batch of soapy water. Then place it back under the lights.

7. Continue doing this until your water trap is flea-free.

What to Buy If You Can't Make Your Own Pest Controls

If you are too worried about making your own pest controls, that's okay, but please try to use natural-ingredient store-bought pest controls if you can. I put together a list of products that use natural ingredients. I for one don't think that there is an absolutely bulletproof listing of "green pesticides." It's an oxymoron, like old news, extensive briefings, random order, and jumbo shrimp. Do not assume that "natural" pesticides are inherently less toxic than synthetic pesticides. Sometimes these products are chemically altered or mixed with other ingredients that are not natural or organic. Thus it's buyer beware. This list of pesticides that are made with natural active ingredients and which trade names that are derived from or sound like natural sources may not be so green. To find where to purchase any of the products on the list, just type the name in an Internet search bar, and hit "Search." Again, buyer beware. Don't assume that a natural-sounding or looking pesticide is also organic or even natural, as well as considered green.

Understanding All Types of Pesticide Risk

No matter how green, natural, or organic, all pesticides have risk to the user's health. What might seem benign or harmless at the time can carry grave consequences. As I always say, even pure clean water is deadly if it's above your nose. Fortunately for us, on the products we purchase that are potentially hazardous, there are labels that state the risks to us as signal words and symbols that are important clues in recognizing how potentially dangerous the product is. All signal words must appear in large letters on the front panel of a pesticide and must say, "Keep Out of Reach of Children," which appears on every pesticide label.

DANGER This word signals that the pesticide is highly toxic. A small amount, if taken by mouth, could kill an average-sized adult. Any product that is highly toxic orally or absorbed via our skin or through inhalation, and that also causes severe eye and skin burning, will be labeled "DANGER." with the word POISON printed in red with a skull and crossbones symbol.

WARNING This word signals that the pesticide is moderately toxic. As little as 1 ounce, taken by mouth, could kill the average-sized adult. Any product that is moderately toxic orally or absorbed via the skin or through inhalation or causes moderate eye and skin irritation will be labeled "WARNING."

CAUTION This word signals that the pesticide is slightly toxic. An ounce or more taken by mouth could kill the average adult. Any product that is slightly toxic orally or absorbed via our skin or through inhalation or causes slight eye and skin irritation will be labeled "CAUTION."

Other precautionary statements are included on pesticide labels to help you to decide the proper steps to take to protect yourself, your family, and other persons and domestic animals that may be exposed. These statements are sometimes listed under the heading, "Hazards to Humans and Domestic Animals." They are composed of several sections, and many are self-explanatory.

Statements that immediately follow the signal word, either on the front or side of the pesticide label, provide you with which route or routes of entry must be particularly protected. Many pesticide products are hazardous by more than one route.

Most "DANGER" statements include

- Fatal if swallowed.
- Poisonous if inhaled.
- Extremely hazardous by skin contact or easily absorbed through skin.
- Corrosive. Can cause eye damage and severe skin burns.

Most "WARNING" statements include

- Harmful or fatal if swallowed.
- Harmful or fatal if absorbed through the skin.
- Harmful or fatal if inhaled.
- Causes severe skin and eye irritation.

Most "CAUTION" statements include

- Harmful if swallowed.
- May be harmful if absorbed through the skin.
- May be harmful if inhaled.
- May irritate eyes, nose, throat, and skin.

TIP **Just because caution statements usually are more moderate and warnings are qualified with "may" or "may be," the lower toxicity levels of products possessing a "CAUTION" label require great respect.**

Route of Entry

Following the "route of entry" statements are the "specific action" statements. They tell us the specific action that should be taken to prevent poisoning accidents. The statements are related to the toxicity of the pesticide product and the route or routes of entry, which must be particularly protected. Example: "Do not breathe vapors or spray mist" and "Avoid contact with skin or clothing" are common specific action statements that help to prevent pesticide poisoning.

Protective Clothing

Pay attention to the "protective clothing" and "equipment" statements that are listed on many pesticides and should be followed closely. Despite the fact that some labels do not contain these statements, this won't mean that protection isn't necessary and you're in the clear. Long-sleeved shirts, long-legged jeans, and gloves should be worn when applying pesticides.

Practical Treatment

The statement of "practical treatment" gives the recommended first-aid treatment in case of accidental poisoning. All "DANGER" and some "WARNING" and "CAUTION" labels contain a note to physicians describing the appropriate medical procedures for poisoning emergencies and may identify an antidote. If you have an emergency, have that note nearby to save time and help those who are performing the first aid or treatment.

Environmental Hazards

The "environmental hazards" statement warns of potential hazards to the environment. Read closely for special warning statements. "Special toxicity" statements warn of potential hazards to wildlife, insects, or aquatic organisms. These statements can help you to choose the safest product for a particular job. "General environmental" statements appear on almost every pesticide label. They are reminders to use common sense to avoid contaminating the environment. "Physical or chemical hazards" statements tell of special fire, explosion, or chemical hazards the product may create. This "classification" statement indicates whether the pesticide is classified as a "general-use" or "restricted-use" pesticide. Just because a product is a general-use pesticide doesn't mean that the product has a low hazard level.

Reentry Statement

Some pesticide labels with the signal word "DANGER" or "WARNING" contain a "reentry" statement. This statement tells how much time must pass before you can reenter a treated area without appropriate protective clothing. These reentry intervals are set by both federal standards and some states, so check with your state to be safe and legal. The reentry statement may be printed under the heading "Reentry," or it may be in a separate section with a title such as "Important," "Note," or "General Information." If no reentry statement appears on the label, then sprays must be dry or dusts must be settled before reentering or allowing others to reenter a treated area without protective clothing. This is the minimum legal and commonsense reentry interval. All pesticide labels contain general instructions for the appropriate storage and disposal of the pesticide and its container. State and local laws vary considerably, so specific instructions usually are not included; if in doubt, check your local and state laws.

Use the Directions

"Directions for Use" is probably the most important part of the label. This is the part of the label that tells you, the consumer, how to use the product. It gives information about the pests the product claims to control; the crop, animal, or site the product is intended to protect; the form in which the product should be applied; the proper equipment to be used; how much to use; mixing directions; compatibility with other often-used products; toxicity; other possible injury or staining problems; and where and when the material should be applied. Additional information includes the least number of days that must pass between the last pesticide application and the harvest of crops, thus allowing time for the pesticide to break down in the environment, which prevents illegal residues on food, feed, or animal products and possible poisoning of grazing animals.

Read and Understand

The important thing is to *read the label*. Read it before you purchase the pesticide, before you mix the pesticide, before you apply the pesticide, and before you store or dispose of the pesticide. Understand what you are reading, and ask questions if you don't. Also included on all pesticide labels is a notice that states that the buyer assumes all responsibility for safety and use not in accordance with directions. Thus, if a salesperson sells you a nonselective herbicide product, and now your entire yard is dead, that's too bad because it's up to you to read and understand the label.

A List of Pesticides That Contain Natural Ingredients

Inorganic Materials or Mineral-Derived Pest Controls

Boric acid and borates are derived from borax, a mineral in the soil. Boric acid and orthoboric acid are essentially the same thing, with orthoboric acid being the proper name. Borates are the same as borax, which contains sodium, boron, and oxygen. Boric acid/orthoboric acid is the more refined crystalline material derived from the borate/borax. In all its forms, boric acid acts as a stomach poison, causing disruption of the proper digestion of food and resulting in starvation (to death).

Active Ingredient

Boric acid, orthoboric acid, pentahydrate borax, sodium *tetra*-borate decahydrate (borax), and/or disodium octaborate tetrahydrate.

Dust Products

Armor Guard, BorActin Insecticide Powder, Boracide Borate Powder, Boric Acid Insecticidal Dust, Borid, Borid Turbo, Mop-Up, Perma-Dust PT-240, Roach Proof Powder, and Victor Roach Powder.

Granular Bait Products

Intice 10 Perimeter Bait, Intice Coarse Granular Bait, Intice Fine Granular Bait, Intice Select Ant Granules, MotherEarth Granular Scatter Bait, Niban FG granular, and Niban granular.

Liquid and Syrup Bait Products

Drax Liquidator, Gourmet Ant Bait Liquid, Intice Thiquid Ant Bait, PT381B Advance Liquid Ant Bait, Terro PCO Liquid, and Uncle Albert's Super Smart Ant Bait.

Gels, Pastes, and Pressurized Applications

388B Advance Ant Gel Bait, Ant Fix Ant Gel Bait, Ants-No-More Ant Bait Gel, Attrax Roach Bait Dual Syringe, Ant-X 75, CB 441 Protein Paste Ant Bait, CB 441 Carbohydrate Gel Ant Bait, Drax Ant Kil Gel, Drax Dual Syringe, Drax NutraBait, Drax Roach Gel, Gourmet Ant Bait Gel, Intice Sweet Ant Gel, Intice Smart Ant Gel, Intice Roach Bait, and Uncle Albert's Super Ant Bait.

Desiccant Pesticides

Insecticides that have a desiccant action on bugs are used in pest control to kill a variety of pests. Desiccants or desiccant insecticides have the ability to dry out bugs and kill them. Not all insects are killed immediately by a desiccant owing to the thickness or hardness of their cuticle or outer shell.

Desiccant Dust/Granule Products

Limestone (Calcium Carbonate)

Limestone in powdered form acts as a desiccant and is a very common sedimentary rock of biochemical origin. It is composed mostly of the mineral calcite. The calcite is derived mostly from the remains of organisms such as clams, brachiopods, bryozoa, crinoids, and corals. These animals live on the bottom of the sea, and when they die, their shells accumulate into piles of shelly debris. This debris then can form beds of limestone. Some limestones may have been derived from nonbiogenic calcite formation. Although some limestones can be nearly pure calcite, there is often a large amount of sand or silt that is included in the shelly debris. Lime is sold in a powdered or pellet form.

Silica Gel

Derived from mineral quartz, sand also acts as a desiccant.

Silica Gel Products

Drione, Diatect II dust, Diatect III dust, Diatect V dust, Tri-Die bulk dust, and Tri-Die pressurized aerosol.

Diatomaceous Earth

This is the by-product of skeletons of oceanic diatoms mined from soil deposits.

Diatomaceous Earth Products

Concern DE, Insecto DE, MotherEarth DE, and MotherEarth Exempt Granules DE.

Miscellaneous Minerals

Clean Kill—silver + citric acid—surface disinfectant; Earthcare Odor Remover Bags—

sodium aluminosilicate, a natural mineral; Earthcare Odor Remover Granules—sodium aluminosilicate, a natural mineral; Earthcare Odor Remover Pouch—sodium aluminosilicate, a natural mineral; Sluggo Snail and Slug Bait—iron phosphate; Pentathlon LF—manganese/zinc—foliar fungicide.

Sulfur and Lime Sulfur

This is an insect deterrent sold in garden centers. Sulfur Dust.

Copper Products

These products are significant because they have fungicidal properties. Bordeaux Mixture combines copper sulfate + calcium hydroxide (lime); Copper Count-N—metallic copper; Kocide 2000—fixed coppers; Kocide DF—fixed coppers; and Nautique Aquatic Herbicide—copper carbonate.

Oils and Soaps (Salts and Fatty Acids)

These are good pest repellents as well as insecticides, and they work by stripping the insect of its protective layers or by suffocating it. M-Pede Concentrate—soap (= safer soap), for use against plant pests.

Microbial Products and Living Organisms

These are basically products derived from living organisms that compete with, feed on, or repel unwanted pests.

Living Pesticides

These are live microorganisms such as *Bacillus thuringiensis israelensis* (BTI). Products include Aquabac 200G—BTI; Avid—abamectin or avermectin, derived from soil fungus, for plant pests (BTI); Briquets—BTI; Dipel Pro—BTI, plant pests; Gnatrol—BTI; Millenium—*Steinernema carpocapsae*, nematodes, ground-dwelling insects; Mosquito Bits—BTI; Mosquito Dunks—BTI; Teknar CG Granules—BTI; Teknar G Granules—BTI; Teknar HP-D Larvacide—BTI; and Vectobac 12AS—BTI.

Botanicals

These are plant-derived products made from parts of plants as well as plant by-products that can help to control pests and can be used as an insecticide.

D-Limonene or Linalool

These are both essential oils extracted from citrus skins. Products include Demize EC—linalool, flea control, synergized with PBO; MotherEarth Wasp & Hornet Jet Spray—limonene only; Orange Guard—D-limonene, general insect control; Orange Guard Ornamental Plants Concentrate—D-limonene; and Power Plant XT-2000 Termiticide for Drywood Termites—D-limonene.

Neem Oil

This is extracted from the Neem fruit tree. Neem oil works as an insect control by entering the insect's system. Once it enters, it begins blocking or replacing the insect's real hormones, making them "forget" to eat or to mate and to stop laying eggs. Some forget that they can even fly. If eggs are produced, they don't hatch, and the larvae don't molt.

Neem Oil Product

Azatrol EC—azadirachtin + liminoids.

Rotenone

This occurs naturally in the roots and stems of several plants and is extracted from roots of tropical legumes. It is a broad-spectrum pesticide that works well against mites and lice.

Sabadilla

This comes from the seeds of the Sabadilla lily, also called *Veratrine*. It is used as a pesticide to prevent insect damage on citrus plants.

Mint Oil

Mint oil is well known for its ability to repel mosquitoes or to kill some common pests such as wasps, hornets, ants, and cockroaches.

Mint Oil Products

Victor Poison Free Ant & Roach aerosol, Victor Poison Free Ant Killer, Victor Poison Free Flying Insect Killer, Victor Poison Free Wasp & Hornet.

Castor Oil

This is extracted from the seeds of the castor plant and is used in many products. The seed hulls are highly toxic because they contain toxic ricin. The oil contains a small amount of this toxin. Castor oil is an effective repellent for moles and as a horticulture oil to discourage insects.

Castor Oil Products

Chase Granular Mole and Gopher repellent, Chase Liquid Mole and Gopher repellent, Mole Med mole repellent.

Pyrethrum

This is an extract from chrysanthemums, a common household flower. It is by far the most common and widely used organic botanical active ingredient in insect-control products. Nearly all pyrethrum products are combined with either one or two synergists that enhance the effectiveness of the pyrethrum. The most common synergist is piperonyl butoxide (PBO), which can make the product no longer organic.

Pyrethrum Aerosols

1600 X-Clude Timed Release, CB-38 Extra, CB-40 Extra, CB-80 Extra, CB-123 Extra, Clean Air Purge I, Clean Air Purge II, Clean Air Purge III, Clear Zone Farm Fly Spray, Directed Spray Aerosols, Konk 1 Flying Insect Killer, Konk BVT Flying Insect Killer, Konk 2 Flying Insect Killer, Microcare Pressurized, P.I. Contact Insecticide, PT-565 Plus XLO Formula 2, PT-580P, and Tri-Die Pressurized—silica + pyrethrum.

Dusters

Drione—combines pyrethrum + silica gel; Tri-Die bulk dust—silica + pyrethrum dust with PBO; Tri-Die pressurized aerosol—silica + pyrethrum dust with PBO.

Essential Tree Oils and Miscellaneous Botanical Oils

This is the proprietary name for a combination of essential tree oils in ecosmart products.

Dusters

EcoExempt D—hexa-hydroxyl + clove oil; EcoPCO D-X—hexa-hydroxyl + pyrethrins.

Granules

EcoExempt G—clove oil + thyme oil; MotherEarth Exempt G, 25 lb—cedar oil + wintergreen oil.

Liquid Concentrates

EcoExempt IC2—hexa-hydroxyl + rosemary oil + peppermint oil; EcoExempt IC—hexa-hydroxyl + rosemary oil; EcoExempt MC—oils of rosemary, cinnamon, lemongrass; EcoPCO EC—hexa-hydroxyl + PBO synergist; EcoPCO EC-X;

Organocide Insecticide/Fungicide—sesame oil, edible fish oil, lecithin.

Wettable Powders

EcoPCO WP-X—hexa-hydroxyl + thyme oil + pyrethrins.

Aerosols

Bioganic Crawling Insect Killer—rosemary oil; Bioganic Flying Insect Killer—eugenol (clove oil) + sesame oil; EcoPCO ACU—hexa-hydroxyl only; EcoPCO AR-X—hexa-hydroxyl + pyrethrins; EcoExempt Jet Wasp & Hornet—hexa-hydroxyl + rosemary oil; EcoPCO Jet-X Wasp & Hornet—hexa-hydroxyl + PBO synergist; EcoExempt KO—hexa-hydroxyl + clove oil.

Herbicides

EcoExempt HC—hexa-hydroxyl + clove oil; Matran EC—clove oil.

Repellents

Critter Ridder Animal Repellent—oil of black pepper, piperine, capsaicin; Ro-Pel Animal Repellent—thyme oil, peppermint oil, white pepper.

CHAPTER 10

Just-for-Fun Projects and 101 Cool Tips

Before you recycle or dispose of anything, you should consider whether it has any life left in it. Start with simple ideas, such as: A jam jar can become a glass from which to drink. Food scraps can be added to a compost pile. An old shirt can become a rag. An outgrown child's red wagon can become a gardening cart. Music CDs can be shared. Old DVDs can be traded in for other ones. Appliances can be repaired or bought used. A computer can be upgraded to perform like new. A car can be rebuilt. Cell phones, ink cartridges, and working electronics can be donated.

Reusing or repurposing keeps new resources from being used for a while longer and old resources from entering the waste stream by finding new life. Making it fun or turning it into a source of income is all the more reason to find new life for old things. I put together some ideas from looking around in my own home and had a lot of fun coming up with creative ways to reuse things.

101 Cool Tips from A to Z

Aluminum Foil

1. Use recycled aluminum foil to quick clean your BBQ grill (see Project 57).

Automotive Floor Mats

2. Use an old floor mat to prevent rust and leaks from metal containers on cement garage or basement floors.
3. Use to put muddy shoes on before entering your house.
4. Place clean mats around litter boxes to keep cat litter from being tracked all over the house.

Billiard Balls

5. Keep the old set in case of mishap and a ball is lost or broken.
6. Keep the old set for practicing those trick shots or for when the kids are playing so that they don't damage or destroy the good ones.

Blankets

7. Keep one in your car for emergencies.

Breath Mint Containers

8. Fill with aspirin or other nonprescription medicines, and keep it in your car or wherever.

Bricks

9. Salvage (with permission) from construction sites and use for edging, or make a walkway around your home and garden (see Project 58).

Briefcase

10. Use it for an overnight bag.
11. Put tools in it, and keep in your car.

Calendars

12. Mat and frame the calendar pictures that you really enjoy.

Candles

13. Melt down all your old candles into one big candle (see Project 59).

Carpet

14. Cut into strips wide enough to fit between the rows in your garden (see Project 60).

Clothes

15. Use for rags.
16. Keep old winter clothes in your car for emergencies.

Clothes Hangers

17. Use plastic clothes hangers to hang wet clothes on a shower curtain rod; when dry, leave on hanger and place in closet (see Project 61).

Dog Food Bags

18. Use as bags for collecting or storing woody yard waste such as sticks and mulch.

Duffel Bag

19. Store out-of-season clothes in it, and then slide it under the bed or in the closet.
20. Fill it with emergency equipment (warm clothes, first aid, etc.), and keep in your car.

Fabric Softener Sheets

21. Use spent dryer sheets on glass shower doors to remove soap and hard-water stains.

Garden Hose

22. Poke small pin holes throughout the length of the hose, plug off the end, and make it into a garden soaker hose (see Project 63).

Glass Beverage Bottles

23. Use old wine and beer bottles as candle holders.
24. Make a wine bottle into a flower vase.
25. Use a larger glass bottle as a bank.
26. Make long-neck bottles into oil candles (see Project 64).

Paper Grocery Bags

27. Use for garbage bags around the house.
28. Put items for composting into paper bags, and compost the bag and all.

Laundry Baskets

29. Use as a recycling basket.
30. Use to store children's toys in a closet.
31. Keep in the trunk of your car to keep groceries together.

Lint (from the Dryer)

32. Use for making a campfire starter with veggie oil (see Project 65).
33. Compost.
34. Place lint in the paths of slugs to discourage them (because the lint gets stuck in their slime, they avoid crossing over it).

Milk Containers (Half-Gallon)

35. Use the bottom half of a half-gallon milk carton to start plants in before moving them to your garden. Use the top half as miniature green houses to protect seedlings from frost.

Newspaper Paper

36. Use the Sunday comics as wrapping paper.
37. Use individual balled-up sheets as packing material.
38. Wrap fragile items for packing.
39. Use for drying windows and mirrors without streaking.
40. Make it into a garden weed block by laying several sheets of newspaper down and spreading mulch on top of them. This will stop most weeds and compost itself for next year's garden.

Paper Egg Cartons

41. Use for potting or starting plants before moving them outdoors (see Project 66).

Pet Fur

42. Throw outside and let the birds use it for their nests.
43. Compost.
44. Place dry fur (or human hair) in the paths of slugs to discourage them (because the fur gets stuck in their slime, they avoid crossing over it).

Pill Bottles

45. Remove or cover over medicine labels when empty so as to not confuse what's in the bottle.
46. Use for safe storage of thumbtacks, small screws, nails, pins, or paperclips.
47. Wash and use for bringing salt, sugar, and pepper along with you in your lunchbox.
48. Fill with aspirin, antacids, or nonprescription decongestants and keep in the car.

Plastic Beverage Bottles

49. Refill with water, juice, or sports drinks and freeze. They make great ice blocks that don't fill your cooler with water as they melt. Plus, you can drink the water as it melts.
50. Make a loud shaker by adding a few coins to an empty bottle, and use it to train your pets. The sound of the coins rattling in the bottle is unpleasant and can be used to deter negative behavior with the negative sound. This is a great way to teach pets not to jump on tables, countertops, and automobiles (see Project 67).
51. Protect your seedlings with 2-liter plastic soda bottle greenhouses. Cut the bottoms out of the bottles, sink them into the soil around the seedlings, and remove the caps. You can reuse them over and over each year.
52. Use to bottle your own bottled water or sun tea (see Project 68).
53. Use large-mouth sports drink bottles for making frozen juice drinks.

Plastic Containers

54. Use for storage of any objects that will fit in them.
55. Save or freeze leftovers in them.
56. Poke holes in a butter container, and use it as a makeshift colander.
57. Use butter containers as cereal bowls or to bring leftovers to work for lunch.
58. Use butter containers as outdoor food and water dishes for pets during the summer.
59. Use yogurt and dip containers for making candle molds (see Project 59).
60. Use fast-food plastic cups when on picnics or outside. Just wash and reuse them over and over.
61. Use a fast-food plastic cup as a scoop for your pet's food.
62. Use them for making sand castles.

Plastic Grocery Bags

63. Use the plastic bags you get from stores for liners in small garbage cans around your house.
64. Keep them in your car for whatever (see Project 69).
65. Make a cat bed from plastic shopping bags (see Project 70).

Plastic Swimming Pools

66. Fill with water for your pet to bath or play in on hot summer days.
67. Use it upside down to protect plants from frost.
68. Poke holes in the bottom for drainage, and fill with soil; then plant a raised-bed garden in it.
69. Poke holes in the bottom and use as a sandbox.
70. Keep on the floor of your utility room, and use for muddy, wet, and dirty clothes for washing.

Post-its

71. Fold over the sticky part and reuse its backside.

Scrap Paper

72. When doing first drafts or printer proofs, print on the reverse side of used photocopier paper.
73. Shredded paper works for packing material.
74. Use for making homemade paper (see Project 71).
75. Make a piñata (see Project 72).

Bed Sheets

76. Cut into pieces, and use for rags.
77. Use as tarps for floors or furniture when painting.
78. Keep one in the trunk of your car, and use it to sit or lie on to work on the car or change a tire.
79. Use to cover your outdoor plants to protect them from frost.
80. Use as a dust cover for furniture.

Shoelaces

81. Save and use for anything for which you would use a heavy-duty string.
82. Use as a replacement drawstring on hoodies or sweatpants.
83. String nuts and washers through for orderly storage.

Shower Curtain

84. Use as an outdoor tarp for grills or the wood pile or as a picnic table cover.
85. Line the trunk of your car with it for easy cleanup.

Tennis Balls

86. Use tennis balls to help you park your car (see Project 73).

Tires

87. Make a tire swing (see Project 74).

Toilet Paper and Paper Towel Tubes

88. Start your seedlings in sections of these cardboard tubes (see Project 75).

Toothbrush

89. Use for cleaning shoes.
90. Use for cleaning threads on bolts and screws.
91. Use for cleaning auto parts.
92. Use for cleaning coffee makers.
93. Use for cleaning the grooves between the tiles in your bathroom.

Child's Wagon

94. Use for gardening and yard work.
95. Use when transporting wood to your fireplace and garbage to the curb.
96. Use for storing your automotive tools, making a rolling toolbox.

Water

97. Use water from a dehumidifier to water your plants.
98. Use your bath water to water your plants.
99. Keep a bucket in the shower to catch water for watering plants.
100. Shower with your plants.
101. Run your clothes washer and dishwasher only when they are full.

Project 57
Use Recycled Aluminum Foil to Quick Clean Your BBQ Grill

Barbequing or cooking food on a grill with heat and burning hot gases from gas, fire, smoking wood, or hot briquettes of charcoal can make for a big burned-on crusty mess. When your grill is dirty, it is unsanitary to cook on. The food that is stuck to your grill is a breeding ground for germs, and burned food can contain carcinogens. In addition to being unhealthy, a dirty grill also can be a safety hazard. The pieces of food that get burned turn into a charcoal-like substance that can flame up. When this happens, it increase the fire hazard of your grill, especially on gas grills. Those pieces also can damage the functionality of your grill and the flavor of your food. I get easily irritated by the sight of a messy grill, greased with oil and soiled with the remains of smoked/cooked/roasted food. Most of us, when it comes time to grill, don't spend much time cleaning the grill from the last grilling, so this is a method (my brother showed me over a decade ago) to quick clean the grill, and it works well between more proper grill cleanings or even when you're using a public grill and don't have soap and running water nearby. Always BBQ outdoors in a well-ventilated area with adequate clearance around the BBQ.

TIP

- Fire is the most destructive force in the universe, and you are inviting it into your backyard for a little cookout. The first thing you need is a fire extinguisher. Next, you need to know your fire and know how to control it. Every year grills and smokers cause thousands of fires, hundreds of injuries, dozens of deaths, and millions of dollars in damage. You need to know how to cut the fuel supply, extinguish a fire, and call the fire department. Always have one person in charge of the fire at all times.
- Read the grill's operator's manual. Follow all the safety instructions. If you have a fire and you didn't follow the instructions, it is your fault. Grills, gas or charcoal, and smokers both have very specific ranges of operation. You need to know these before you light up.
- Finding the proper spot to locate the grill is very important. Your owner's manual will tell you the minimum distances around the grill that must be kept clear. You also don't want people to walk too close to the grill's hot surfaces, and you want to ensure that children won't be playing nearby.
- Inspect your gas grill for problems such as bugs that climb into little places causing gas to flow where it shouldn't. Gas grills also can produce a great deal of heat and melt through hoses, knobs, and other parts, so look it over carefully.
- Charcoal grills cause more fires than gas grills because of lighting the charcoal. Lighter fluid causes all kinds of problems. Never even think about adding lighter fluid to hot coals. While liquid lighter fluid burns, evaporated lighter fluid explodes. Follow the lighter's instructions exactly, or use another system of lighting your charcoal grill.
- Do not use your BBQ in windy conditions when there is a risk the burners may blow out.
- Remove all excess fat and residue from the BBQ after each use to prevent a fire hazard.
- Regularly check your gas cylinder valves, hoses, and connections for any leaks or blockages. To check for gas leaks, spray a solution of soapy water onto hoses and connections. Bubbles will form if there is a gas leak. Test the connection to the gas cylinder every time it is reconnected.
- Liquid petroleum (LP) gas cylinders should be stored outdoors in a well-ventilated area, upright, and away from heat sources.
- Gas cylinders are required to be tested and stamped every 10 years.
- Always use the correct BBQ gas. Auto gas differs from BBQ gas and should never be used with your BBQ.
- Never attempt to repair gas appliances.
- Ensure that a licensed gas fitter carries out any work (note that not all plumbers are licensed gas fitters).

WHAT YOU'LL NEED

- A dirty grill
- Recycled aluminum foil
- Heat-resistant gloves
- A long BBQ spatula (Figure 10-1)

Figure 10-1

Let's Start

1. On a cool grill surface, take a wad of your recycled foil and scrub your grill surface with it to loosen up as much stuff stuck to it as you can. If you have soap and water, wipe the surface down if you can to remove even more gunk.
2. Next, fire up the grill, and let the grill surface get hot for about 15 minutes.
3. Now carefully, with your glove on, take another wad of foil and wrap it over your spatula and use it again to scrub off any ash or loosened grim that the heat created.
4. With the ash and residue gone, plus with the germ killing heat, go ahead and use the grill.

Project 58
Making a Walkway from Used Brick Pieces

Many builders use brick for the exteriors of the houses and other structures they build. The process of laying brick can create a fair degree of waste from fitting and cutting the bricks to size. Most often these pieces are used to fill in voids around foundations or are just tossed into the trash. By going to building sites and collecting these pieces, you could have enough to build a unique garden path for little to no money. When you are done building your walkway, you will have a unique, beautiful, and simple path that will add a lot of character to your home.

WHAT YOU'LL NEED

- A square-tip shovel
- Brick pieces
- Bags of cement and gravel premix
- A rake
- Gloves
- Old newspaper
- A sledgehammer
- Eye-protection goggles
- Water
- A water hose and sprayer
- A hard-bristled broom (Figure 10-2)

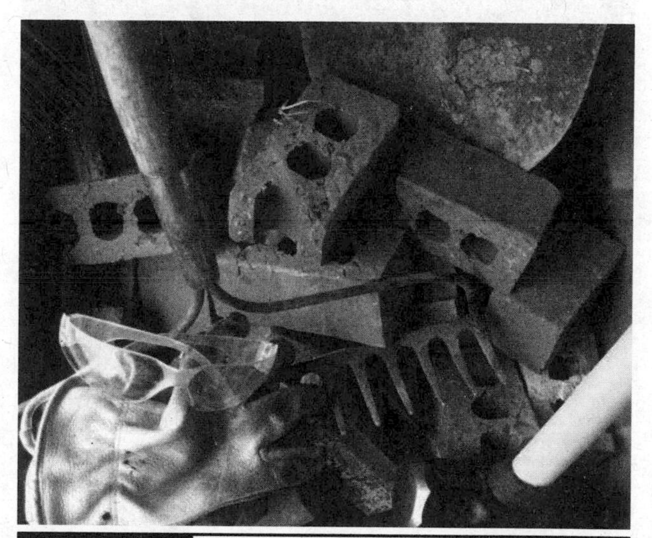

Figure 10-2

Let's Start

1. You will need a lot of brick pieces to make a walkway, so keep this in mind before you start. The first thing you will need to do to create your walkway is to plan out where you want it to go. You can either make your walkway straight, or you can make it more winding and casual. Use flour or sticks to mark out the two sides of your walkway. Be sure to make your path wide enough for two people to walk side by side.
2. Next, you will need to remove any grass or weeds that are in the path. Use a square-tip shovel. If you use a shovel, hold it as close to flat to the ground as you can, and work it underneath the sod to remove the sod in chunks instead of actually trying to dig it up.

3. Next, place newspaper on the path, overlapping it often and making it over 10 sheets deep (wet the paper if it's windy). The newsprint will help to prevent weeds from growing up in your pathway after you have spread the brick pieces (Figure 10-3).
4. Finally, start placing the brick pieces in the path, with any sharp or broken edges facing down. And for the overly large pieces, use the sledgehammer to break them up. Spread the brick pieces out into the path with a shovel or rake to smooth the path out (Figure 10-4).

Figure 10-3

Figure 10-4

5. Spread the cement and gravel premix (dry) between the brick pieces, and sweep it down between the brick pieces, exposing the top surfaces.
6. Once the spaces between the bricks are filled with the cement and gravel premix, carefully and slowly sprinkle water over the path. The cement won't be what holds the path together, but it will keep the bricks from moving excessively. As the cement breaks apart in time, it cradles the brick pieces (Figure 10-5).

Figure 10-5

Project 59
Making Old Candles into New

Making candles at home was a basic requirement of everyday life for many people fewer than a hundred years ago. Once the sun went down and before the light bulb was invented, life went on around the glow of candles. Candles continue to be made at home, but no longer out of necessity for lighting the night but because candle making is an activity some people like to do. Even store-bought candles are common in our homes because they are inviting, add a sense of warmth, and can create a pleasant odor to hide or remove other less pleasant odors. When we use candles, homemade or store-bought, it never fails that we end up with unused portions of candles left over. These pieces can be recycled into another candle easily with a few things you already may have in your kitchen.

WHAT YOU'LL NEED

- An electric crock pot
- Wax candles
- Wax
- Crayons (optional; use sparingly)
- Small polypropylene (PP #5) plastic containers (yogurt and dip type)
- Small glass containers (recycled glass votive candle holders)
- A freezer
- Water
- A hammer
- A screwdriver
- Candle fragrance oil (hobby/craft stores)
- Candle wick kit (hobby/craft stores)
- Toothpicks
- A razor knife
- Gloves (Figure 10-6)

Figure 10-6

TIP

- **Buy candles at garage sales and after-holiday clearance sales; these are good sources of cheap candle feedstock wax.**
- **Never leave a burning candle unattended.**
- **Keep candles out of the reach of children and pets.**
- **Always burn candles on a heat/fire-resistant surface. Even if your candle is set on a heat/fire-resistant plate or candle holder, be sure that the candle holder itself is placed on a safe, stable surface.**
- **Always keep candles well away from flammable materials such as curtains, lampshades, decorative items, plants, and bedding.**
- **Trim your wick to about ¼-inch every time you burn a candle.**
- **Keep candles out of drafty locations.**
- **Keep wick trimmings and other debris out of the pool of melted wax.**
- **Never move a lighted candle.**
- **Don't burn candles all the way down.**
- **If a candle is not burning properly—say, it is sputtering or smoking or the flame is burning extremely high—don't use it.**

Let's Start

1. Collect all your old candles and broken crayons, and remove any paper, plastic, or anything that might burn.
2. Freeze the candles to make them brittle. Use the screwdriver to carefully remove the wax from glass containers. Then break the larger wax pieces into smaller ones on a hard surface using your hammer. Remove any remaining wick or other flammable pieces at this point (Figure 10-7).
3. Now fill the containers you're going to use as your candle molds. Glass containers are going to house your finished candles.

Figure 10-7

4. Place your wax-filled containers into the empty crock pot, and add water around them up to about two-thirds of the way to the top. Be careful not to get water into your molds. You also don't want to overfill the crock pot with water, making the molds float and tip over (Figure 10-8).
5. Turn the crock pot on to the lowest setting, and give it plenty of time for the wax to melt in the molds. Once the wax is melted, you will want to work quickly before the wax has time to cool. Now you can add the scented oils if you want.

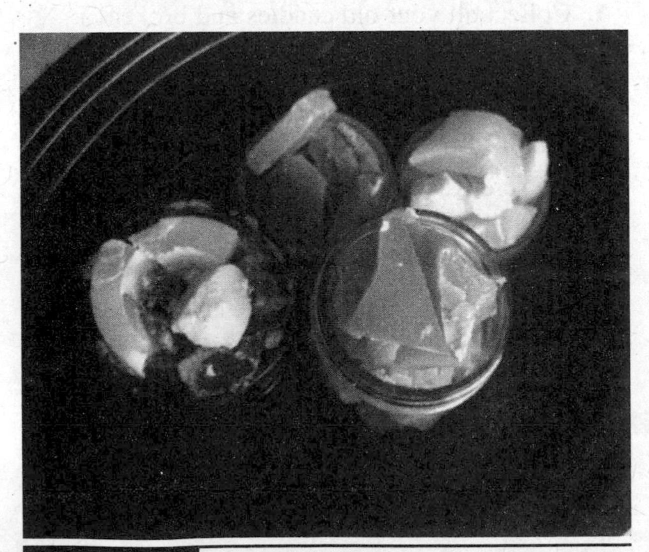

Figure 10-8

6. Now comes time for your wick. Secure the wick by wrapping its top around a toothpick to hold it in place. Now carefully insert the wick into the molten wax, and let the toothpick sit across the container's top holding up the wick. Turn off the crock pot (Figure 10-9).
7. Let the crock pot cool, and remove the molds. Trim the wicks to about half a fingertip's height (Figure 10-10).

Figure 10-9

Figure 10-10

Project 60
Making a Weed Block with Recycled Carpet

Carpet can be used in the best of gardens, although you probably wouldn't recognize it unless you knew it was there. Don't break your back digging and pulling those weeds. Lay carpet down, and top it with mulch or gravel, or leave it plain and uncovered. This will minimize weeds, but keep in mind that this also can be a home for slugs. Therefore, lift the carpet once in a while to remove any pests you may see. Make sure that it is woven-backed carpet and not rubber backed. You want water to pass through it, just not the weeds.

WHAT YOU'LL NEED
■ Hard-toothed rake
■ A razor knife
■ Recycled carpet

Let's Start

1. Make sure that the area is smooth, and rake the area to get a nice even surface.
2. Unroll the carpet lengthwise across the area, and cut it to length. Trim the edges to follow the curves of your garden beds.
3. If you have existing landscape plants, cut a slit to allow the main stem to fit through. Once you slide the carpet into place, enlarge the hole if needed to reduce pressure on the plant stem or tree trunk.
4. If you are planting new plants, install the carpet by first cutting an "X" in it to make the space for planting. Dig your hole, and plant as you normally would. Just make sure that the carpet stays on top of the soil and is tucked away from the emerging new plant.

Project 61
Clothes Hanger

You can recycle plastic clothes hangers from stores. Most department stores have no way to recycle those plastic hangers and are more than happy to give you as many as you want. The flat hangers that clip are great for pants because you can dry the pants and store them on the same hangers. For children, you want smaller hangers, so go to the children's department for those. Don't be shy. You are recycling hangers that are often just wasted. The nice part of doing the drying on hangers is that it makes clothes easier to sort, and you don't need to remove them from the hangers when they are dry. You just hang them all up in the closet (except for the socks and underwear).

WHAT YOU'LL NEED
■ Plastic clothes hangers (metal ones rust)
■ Shower curtain rod or dryer rack
■ A fan

Let's Start

1. During cold winter months, hang as many items as you can on the clothes hangers, beginning with the obvious things, such as dresses, dress shirts, and blouses, and hang the hangers on the shower rod in your shower to dry.
2. Space the clothes so that air can get to them, or the clothes will not dry. You also can hang things such as pajama tops, T-shirts, small kids' shirts, and one-piece outfits. Lightweight pants, pajama bottoms, skirts, and sweats can be pinned on clothes hangers, and even sheets can be folded and hung on them.
3. In the summer months, buy a drying rack, and relocate it outdoors.

Project 62
Build a Recycled CD Scarecrow

The best way to keep birds from your fruit trees is to scare them away or provide other alternatives. Most birds dislike flashes of light because they can help to mask predators from their sight and confuse their escape routes. Recycled CDs catch and reflect sunlight as they spin in the tree and project the light in a random fashion. This is very effective on sunny days and less effective on days with cloud cover.

WHAT YOU'LL NEED
■ Recycled CDs
■ Fishing line
■ A ladder

Let's Start

1. Place an arm's length of fishing line through the center hole of each recycled CD.
2. Tie the ends together so that you can slip tree limbs through the middle of the looped string.
3. Position the string loops with the CDs hanging down over the tree limbs so that they swing in the breeze in a random pattern.

Project 63
Recycling a Garden Hose into a Soaker Hose

A homemade soaker hose can be a specially designed recycled garden hose that you have permeated with tiny holes. Water seeps out through the holes steadily and irrigates the garden by delivering water directly to the roots of plants. You also lose less water to evaporation or runoff, and you'll never accidentally water the driveway or the side of the house, as with a sprinkler. Your plants will be healthier because they are receiving water more directly at the deep root level, where they absorb it more effectively anyway.

WHAT YOU'LL NEED
■ A recycled garden hose
■ An ice pick
■ An off and on spray

Let's Start

1. Place the soaker hose through the garden to get water near to each plant. The water will spread underground, so only 50 percent of the root zone needs to get wet. On slopes, run the hose across the slope, not up and down.
2. At the end of your garden hose, add the on and of sprayer with it in the off position. You use this for pulling off water for watering with a bucket or watering can.
3. Go back to each plant or area you want to water, and carefully push your ice pick into the parts of the hose that you want water to leak out of. To help with restriction and flow, add more holes for each plant as you travel further from the point where water enters the hose. There's less pressure at the end of the hose, so you'll need more holes for the flow to be better balanced.
4. Hook the hose to a water source, and slowly open the valve to let the water just seep (so that it doesn't spray out of the holes).
5. Run your soaker hose for about 20 to 40 minutes once a week. The key is to wet the soil in the root zone of plants. You can tell if the plant is getting enough water by digging a hole with a trowel to see if the root zone is wet (give the water time to soak in). Leaves may droop a little on hot days, but if they stay

droopy after the hottest part of the day, they're probably too dry, so water a little longer or more often. Most plants don't need extra water once they're established. Drought-tolerant or low-water-use plants need watering only for the first 1 to 3 years while their root systems develop. Soaker hoses work well during this initial establishment period. The key is to water deeply but infrequently to develop deep, strong roots.

Project 64
Make Outdoor Oil Candles from Recycled Glass Bottles

Making an oil lamp is very easy, quick, and cheap. The basic elements are nothing more than a few metal washers, a wine bottle, marbles, a length of cotton twine, some vegetable oil, and a vessel to hold it all in. The oil fuel is vegetable oil because it's cheap and works best outside. To use your candle's flame to discourage outside pests, go to a health or hobby store to find "essential" oils from plants such as cedar, lemongrass, rosemary, chamomile, citronella, and eucalyptus that will repel bugs. It will take a good teaspoon per cup of oil to ward off the bugs. You can mix the oils in any combination to make your own fragrance. Be a little reserved, though; the smells can be a little overwhelming when the oil is burned.

TIP **Keep candles at least 12 inches away from anything that can burn. Use candle holders that are sturdy and won't tip over easily on a sturdy, uncluttered surface. Light candles carefully, and keep your hair and any loose clothing away from the flame. Never use a candle if medical oxygen is used nearby. Always have fire extinguishers on hand. Don't let water get on hot glass because it can crack and break into small pieces, spill oil, and make a fire.**

WHAT YOU'LL NEED

- Recycled long-necked glass bottles
- Metal washers, one large enough to cover over the bottle's opening and another one that sits on top of the first washer that has a hole in the center that is a snug fit for the wick you're going to use
- Olive oil or vegetable oil
- Sand
- A razor knife
- A funnel (Figure 10-11)

Figure 10-11

Let's Start

1. Fill a clean and dry long-necked bottle (in this case a wine bottle) over half full of gravel to reduce the amount of oil used in the bottle later.
2. Cut a hand-long length of wick material.
3. Thread the material through the smallest washer (bend it if need be to help make it a snug fit), and have a fingernail's worth piece of wick pulled out of one side (Figure 10-12).
4. Add the vegetable oil fuel and fragrances to the bottle, and leave some airspace from the top of the bottle to the oil's surface.

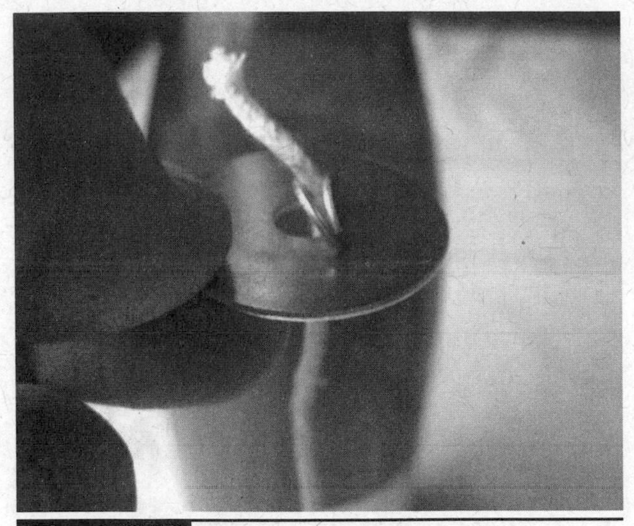

Figure 10-12

5. Take the long end of your wick and thread it through the large washer that's going to sit on top of the bottle.
6. Place the wick inside the bottle, immersing it in the oil, and let the washers rest on top of the bottle.
7. Allow the oil to soak to the top of the wick, and you now can light it.
8. As the wick is slowly consumed, just extinguish the flame and use tweezers to carefully pull more wick out from the washers and relight (Figure 10-13).
9. Add more vegetable oil fuel and wick as needed.

Figure 10-13

Project 65
Making a Campfire Starter from Lint

A campfire is a necessary part of camping. It lights up the campsite at night, boils water, cooks our food, and is also fun to gather around and tell spooky stories. The first thing you will need for a campfire is a flame source (matches or a lighter), followed by some sort of tinder or fire starter. A fire starter should light easily and burn well, and you should always keep it in a watertight, airtight container. Always protect fire starters from moisture so that they work properly and are less likely to be ignited accidentally.

WHAT YOU'LL NEED

- Dryer lint
- Vegetable oil
- A tablespoon
- A resealable airtight container

TIP

- **Only start a campfire in a fire pit or fire ring that is of solid construction.**
- **Avoid starting a fire underneath low-hanging branches or shrubbery.**
- **Don't stack extra firewood too close. If you've recently gathered some, store it upwind so that sparks don't fly into your pile.**
- **Don't allow children and pets near the campfire, and never leave them unsupervised.**
- **Teach kids how to stop, drop, and roll if their clothing catches fire. Have a fire extinguisher handy for emergencies.**
- **Keep your fire away from anything flammable, such as dry grass, tents, paper plates and napkins, and camping gear.**

- Be aware that hot embers can reignite the fire if strong winds are blowing. Shuffle the fire, and make sure that it's out before going to sleep.
- Always have on hand things to put out the fire, such as water, a shovel, and a fire extinguisher, and make sure that the fire is completely out before leaving it unattended.

Let's Start

1. Fill the container you are going to use to store your fire starter with your dryer lint until it's packed tight and near the top of the container.
2. Cover the top of the lint with a layer of vegetable oil, and let it soak in for a few minutes.
3. Using the tablespoon, move the lint around, exposing a dry area, and soak it with a layer of vegetable oil. Repeat this until you see no more dry lint material.
4. Seal the container, and it's ready to use.

How to Use It

1. Take out a handful-size piece of the starter, and place it under some small twigs.
2. You now can ignite your fire starter with a lighter, matches, or a magnifying glass.
3. Gradually add larger sticks. Be careful not to quickly add larger pieces of wood into the fire; that could result in sending burning embers into the air.
4. Use the least amount of wood necessary to keep the fire going to make it easier to put it out later on.

Project 66
Paper Egg Carton Planter

Recycling paper egg cartons and using them for jump-starting seedlings for planting is a great way to put them to good use. You can even leave your seedlings in the paper from the egg carton when you plant them because the paper is going to biodegrade once it's in the soil.

WHAT YOU'LL NEED
■ Paper egg cartons
■ Waterproof tray
■ Potting soil
■ Seeds
■ Water
■ A razor knife (Figure 10-14)

Figure 10-14

Let's Start

1. Start by carefully removing the tops of the paper egg cartons with your razor knife, and add them to your compost pile or recycle them with your cardboard.

2. Cut a slit across the bottom of each egg cup.
3. Now place the carton in your waterproof tray, and soak the carton with water.
4. Fill the pots to within ¼ inch of the top with potting mixture, and level the surface. It's also good to water the soil and allow it to drain thoroughly before sowing the seeds. Make a hole for each seed with your finger or a pencil. Keep in mind that most seeds need to be planted four times as deep as the seed is wide. If your seeds are very small, cover them with a light layer of soil (Figure 10-15).
5. Germinating soil should be kept moist but not soaking wet. Too much moisture will cause the seeds to rot, so use your tray to water from the bottom up. If you can, slip your tray into a clear plastic bag (dry cleaning bag) to keep the humidity and moisture even and reduce the risk of overwatering (Figure 10-16).
6. Once you're ready for planting, separate the egg cups (pots) with your razor knife, and plant the pot and seedling all together.

TIP **Plants like a southern exposure. If you don't have a window that will do, try using growing lights.**

Figure 10-15

Figure 10-16

How to Use It

1. Most seeds require light to germinate, whereas others may prefer darkness. Your seed packet should tell you what your seed's requirements are. Once germinated, all seedlings need light to develop into strong, healthy plants. Supplement the natural light with grow lights if necessary.
2. The care you give your seedlings in the weeks following germination is critical. Keep them moist but not dripping. Small pots and flats can dry out quickly. The first two leaves you will see on the seedling are not true leaves yet, just food storage cells called cotyledons. Once the first true leaves have developed, you can start fertilizing. Choose a good liquid organic fertilizer, or make your own and use it as a weak solution once a week.
3. One week before transplanting your seedlings outdoors, start to harden them off. This process acclimates the plants to get them ready for the wind, cool temperatures, and strong sun in their new environment. Move the plants to a shady outdoor area at first, and bring them indoors for the night if night

temperatures are cold. Each day, move them out into the sun for a few hours, increasing the time spent in the sun each day.

4. Don't set your plants in the garden if they won't withstand frost. Be sure that all danger of frost has passed before setting them out. Water the ground outside and the seedlings thoroughly before transplanting. This helps to prevent transplant shock. It's preferable to transplant on a cloudy day so that strong sun won't wilt your seedlings. Dig a hole about twice the size of the pot, and set the transplant into the hole so that the pot will be covered by ¼ inch of soil. Press the soil firmly around the pot. A small depression around the plant stem will help to trap moisture. Water deeply so that your plants won't develop shallow roots, and water every day for the first week.

Project 67
Make a Dog Training Device

When I was growing up and we would have a new pet to train, we made these shakers that we would use to scare and train our pets, and they worked well and were fun for us kids to use.

WHAT YOU'LL NEED
■ Any type of plastic bottle with a screw-on cap
■ Anything solid and small enough to go into the bottle (change, rocks, nails, etc.)
■ A dog or cat

Let's Start

1. Fill the plastic bottle with just a few pieces of whatever you are going to use, close the cap, and give the bottle a shake.
2. Make several of these bottles that look the same, and place them in areas where they are easy to get to, as well as in places where your pet can see them and avoid going near them.

How to Use It

Animals don't like loud noises, and for some reason, this bottle shaker is a noise they hate. But it's not so loud as to give the pet a noise phobia; that is just being cruel. Just give the bottle a quick shake with your pet watching so your pet can get an idea that you have it first, and then set the shaker down. You are going to borrow a page from what a mother dog does with her litter. Much like a mother dog's low growl, you can make a sound that encourages the dog to rethink its decision. A mother dog would make a low growl that is enough to persuade a pup to think about what it is about to do. You know your dog well enough to read its body language. This is when you want to interrupt the dog with a shake. It just reminds the dog that this is something that you don't want to see happen. This is just another small behavior that establishes you as the leader of this small pack.

For cats, just shake the bottle when they do something wrong, such as jump on a counter or try to defecate in an inappropriate spot. The shaker can correct this behavior, but you have to do it while the cat is in the act. Place the shakers where you don't want the cat to go, such as on a countertop or in a potted plant. They don't even like the sight of the shaker.

Project 68
Make Sun Tea in a Recycled Plastic Bottle

Using recycled plastic PET (1) bottles to make sun tea is easy. All you need is water, tea bags, and the help of the sun.

WHAT YOU'LL NEED

- Clear plastic PET (1) bottles
- Fresh clean water
- Tea bags

TIP There's some concern that brewing tea in the sun can harbor bacteria, and this is a possibility. This is so because the water will get warm enough to provide a friendly environment for bacteria but not hot enough to kill them. To minimize this risk, use a perfectly clean container (scrub it with soap and hot water, and rinse it well), and don't leave the tea to sit unrefrigerated for more than a few hours. Make just enough tea for one day.

Let's Start

1. Try about two tea bags per quart (liter) of water. Place the tea bags in the container, and fill it with cold water.
2. Place the bags' strings under the cap, and screw it down to hold them tightly under the lid.
3. Set your plastic bottle in a sunny place, and let nature do the brewing.
5. When the tea looks and tastes ready, bring it in and take out the bags. Squeeze the bags before discarding them to release all the flavor into the tea.
6. Store the container in the refrigerator, and serve the tea over ice.

Project 69
Make a Plastic Bag Dispenser for Your Car

Keeping old plastic grocery bags in the car is good for when you need them. They have so many uses and come in handy over and over again. Having a way to store and dispense the bags in your car is necessary so that they cannot get free and become litter inside or outside your car.

WHAT YOU'LL NEED

- Recycled plastic grocery bags
- A recycled plastic butter container with a lid
- A razor knife

Let's Start

1. With the razor knife, carefully cut an X across the top half of the butter container's lid.
2. Fill the tub with clean plastic grocery bags.
3. Put the lid back on the tub, and pull out a bag whenever you need it through the X you just cut in the tub's lid. The X in the lid keeps the bags from getting caught in the wind and blowing free.

Project 70
Make a Cat Bed from Plastic Grocery Bags

Our cat loves to lie on and play in the plastic grocery bags I save for recycling. She loves how the bags feel and keep her warm, and she loves the crinkly sound they make for some reason.

WHAT YOU'LL NEED

- Plastic shopping bags (thin film type)
- A recycled pillow case
- A needle and thread

Let's Start

1. Fill the pillow case with the bags until it's about a third of the way full.
2. Use your needle and thread to sew the end closed.
3. Place this in the cat's favorite spot, and watch the kitty go.

Project 71
Make Homemade Paper from Recycled Paper

Making handmade paper can be fun, and it appeals to the need to conserve resources and recycle materials. Higher-end stationery (beautiful and expensive), wrapping paper, and greeting cards are all part of the fashionable trend for paper made by hand rather than by machine. The raw materials for papermaking are free and widely available from your own home. Waste paper can be recycled to make new paper, and we all know how much paper is thrown away around us every day. Printing inks are washed out during the papermaking process. Coarse stems and leaves and plant parts leave particles of fiber in the paper, which add to its natural "handmade" appeal.

WHAT YOU'LL NEED

- Paper
- Water
- A large bowl
- A blender
- A cotton bath/beach towel
- A window screen
- Rags

Let's Start

1. Tear the paper to be recycled into small bits no bigger than 1 × 1 inch, and put in the large bowl.
2. Soak the paper with hot water, and let it sit for several hours to soften the fibers.
3. Now add a handful of wet paper into the blender, and add a enough water to fill the blender to about the halfway point.
4. Cover the blender, and pulp the paper.
5. When ready, pour the pulp out over the window screen until it is about the size of the paper you want to make. Keep adding more pulp to the screen until you reach the desired paper size. A thick layer of pulp will create a thick sheet of paper. A thinner layer of pulp will create a smoother and more delicate sheet. Add some pressure to the pulp to remove some of the water (dewatering the pulp).
6. Now the pulp should be dewatered even further by sandwiching the screen between your hands and a flat surface. The more pressure that is applied, the thinner and smoother the paper will be.
7. After pressing the pulp, flip the screen onto the cotton towel, and tap on the back of the screen, releasing the sheet of paper.
8. Leave the recycled paper to dry on the towel.

Project 72
Make a Piñata from Recycled Paper

Piñatas are used at birthday parties. You can make them in all kinds of shapes and sizes. Piñatas are usually hung in the air, and each blindfolded child take a turn swinging at the piñata, trying to break it open so that the candy and toys inside can spill out.

TIP This is a 3- to 4-day project because of drying time.

WHAT YOU'LL NEED

- An extra-big round balloon or even a beach ball
- A pile of old newspapers
- White flour and water to make a paste
- Scissors
- Masking tape
- Poster board or thin cardboard
- String or twine
- A razor knife
- A saucepan
- A 1-cup (250-milliliter) measuring cup
- Bowls
- A fork or wire whisk
- A stovetop

Let's Start

1. Make your glue with white floor and water.
2. Boil 5 cups of water in a saucepan.
3. In a bowl, mix ¼ cup of white flour with 1 cup of cold water. Mix or whisk it to a smooth consistency (like gravy), completely free of lumps.
4. Now add the mixture to the boiling water in the saucepan.
5. Gently return the mix to a boil while stirring constantly for 2 or 3 minutes until the mixture thickens again.
6. Remove the saucepan from the heat.
7. Cover the glue when it is not in use.

Make the Piñata

1. Spread newspapers over the area where you're going to make the piñata.
2. Air the balloon or beach ball up, and tie a knot at the end.
3. Tear the newspapers into strips about 1 inch wide and about 6 inches long. Tearing rather than cutting is important; it exposes more loose paper fibers and helps the strips lie flat on top of one another.
4. Dip the newspaper strips into the floor glue, and carefully spread them onto the balloon or beach ball, leaving a small hole at the top to remove the balloon and fill the piñata later after it dries. Thoroughly cover the balloon or beach ball with just one layer of newpaper, and let it dry before you add more layers.
5. Once the first layer is dry, keep adding layers of newspaper that is wet with glue until the papier-mâché is built up to a good thickness.
6. The next day, if the paper feels dry when you touch it, then wrap the balloon with a thick string or twine to give it strength to hold when swinging it and the kids are trying to bust it open. Add another two layers of glue and newspapers. Let it dry for another day.
7. Remove the balloon or beach ball. Then go ahead and paint the piñata and dress it up however you see fit. Glue cardboard tubes on it for legs. Make smaller piñatas to use as other body parts. Make streamers from

narrowly cut colorful pieces of paper such as the funny pages or old gift wrapping paper.

8. Fill the piñata with goodies that include candy, of course, and other small items such as balls, toy cars, and so on—and anything else that crosses your mind.

How to Use It

TIP **All other children (while a piñata hitter is in play) must be kept a minimum of a six-step radius from the hitter. In addition, before you allow the other children to grab the goodies spilled on floor, make sure that the hitting of the piñata has stopped.**

1. Collect everything needed to play the game, such as a rope; a place to suspend the piñata, such as a tree branch, a basketball hoop, or even two adults holding a rope with the piñata hung in the middle; a piñata buster stick; and blindfolds for the kiddies. Then hang the piñata.
2. Organize the kids, starting with the smallest ones first and the tallest ones last. If a child is very young, he or she won't require a blindfold. Blindfold older kids, spin them around a few times, provide them the piñata buster stick, and let them loose to play the party game by hitting the piñata. If you are having an indoor party or have very small children, you may prefer an indoor pull-the-piñata version of the game (same rules but just pulled around on the floor).
3. Allow each child to hit the piñata at least a couple of times before you move on to the next child. To make the game even more fun, the piñata must be swung up and down, and guests also should misdirect the piñata hitter to give every kid a chance to play.
4. Keep some of the prize candy nearby to give to the little kids who can't get to the spilled goods so easy, just to make sure that all the kids have fun.

TIP **In my house, I have two piñatas. One for big kids, and one for the little ones. This works well to keep the peace.**

Project 73
Make a Device to Help You Park in the Garage

Use a recycled tennis ball and a long string hanging from the garage door rails to where it touches a planned spot on your automobile's windshield indicating where to stop the car on entering, saving your car and garage from damage and making it easier to get the car parked in the most efficient use of garage space.

WHAT YOU'LL NEED
■ A ladder
■ A heavy string
■ A tennis ball
■ A car
■ A razor knife
■ A strong magnet

Let's Start

1. Park your car in the garage where you want it.
2. Carefully cut slits in the tennis ball, and thread the string through it and tie it off.
3. Using your ladder, hang the ball to where it touches your side-view mirror or another place on the car that you can see easily from the driver's seat.

4. Wrap the string around the magnet, and stick the magnet to the outside of the rail where the garage door rollers don't pass.
5. Just line up the spot on your car with the ball as close as you can whenever you park the car.

Project 74
Make a Better Tire Swing

I had a tire swing when I was growing up, and I spent many hours playing in it, on it, and around it. There is more to building a tire swing than tying your spare to a tree limb with a rope, however.

WHAT YOU'LL NEED
■ A clean tire in good shape
■ A drill
■ Drill bits
■ Good heavy rope
■ A good tree limb
■ A ladder

Let's Start

1. It all starts with locating a properly sized and sufficiently strong tree limb or a wooden frame from which to hang the swing.
2. You need to think about the cleanliness of the tire, so clean the tire completely. Wash the tire at a high-pressure car wash, and then scrub it down by hand using a heavy-duty cleaner. Make sure that you choose a tire that does not have steel belts showing that can result in scratches, cuts, and pain.
3. Drill a few holes through the portion of the tire that will form the bottom of the swing. You don't want water to gather within the tire after it rains.
4. You want to use a large-diameter rope. Select a very thick rope that is obviously stronger than what you probably will need for safety and durability. Make sure that you purchase enough rope. Determine the length needed by adding about 10 feet (3 meters) to the distance from the limb to where the tire will hang.
5. Knot the rope with a common square knot, and make sure that the knot is incredibly strong and is doubled up. Never rely on just a single knot. Test the swing as much as possible to ensure its strength before allowing a child to play on it. If the swing won't support you (an adult), it's not safe enough for kids.

Project 75
Make Seedling Starters from Toilet Paper Tubes

Grow your garden seedlings in these tubes by cutting them into sections and filling with potting soil. Just add plant seeds and water. When they sprout, just plant the compostable tube, seed, and all in the ground.

WHAT YOU'LL NEED
■ Paper towel and toilet paper rolls
■ A waterproof tray
■ Potting soil
■ Seeds
■ Water
■ A razor knife (Figure 10-17)

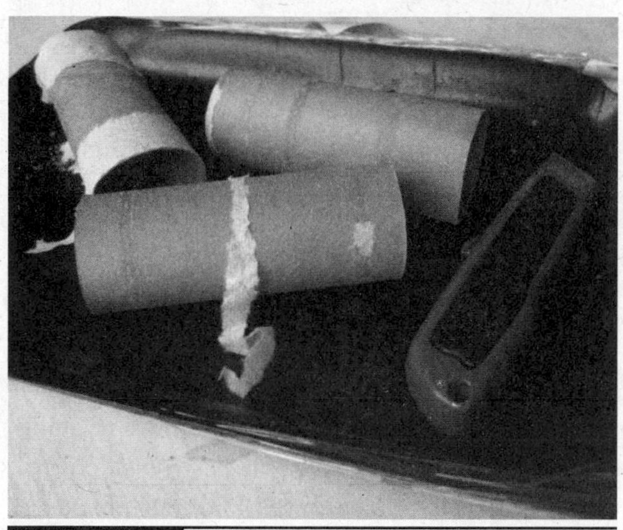

Figure 10-17

Let's Start

1. Cut the tubes into finger-long sections.
2. Place the tubes side by side in the waterproof tray, and add water to the tray.
3. Carefully fill the tubes to within ¼ inch of the top with your potting mixture, and level the surface. It's also good to water the soil and allow it to drain thoroughly before sowing the seeds. Make a hole for each seed with your finger or a pencil. Keep in mind that most seeds need to be planted four times as deep as the seed is wide. If your seeds are very small, cover them with a light layer of soil (Figure 10-18).

Figure 10-18

4. Germinating soil should be kept moist but not soaking wet. Too much moisture will cause the seeds to rot, so use your tray to water from the bottom up. If you can, slip your tray into a clear plastic bag (dry cleaning bag) to keep the humidity and moisture even and reduce the risk of watering (Figure 10-19).
5. Once you're ready, plant the tube and seedling all together.
6. Follow the "How to Use It" directions in Project 66.

Figure 10-19

Project 76
Build a Solar-Powered Hydrogen Firecracker

Build your own hydrogen gas–powered blast cannon or firecracker that's fueled by water and electricity to frighten away pests, celebrate the Fourth of July or New Year's Eve, or frighten birds away. This is a project that's near and dear to my heart because it's such a great example for young minds to learn so much about energy, electricity, chemicals, and the environment. In fact, when I was much younger, I spent many hours experimenting with hydrogen gas and extracting it from water. My fascination with electrolysis started completely by accident. I was charging a lead-acid battery on a workbench, and I was getting ready to test it. I had caused an electric spark as I was removing the charging terminals. In a blink of the eye, the battery exploded, and the top of the battery flew off and hit me in the head and ear. So please be careful around charging batteries, and always turn off the charger when hooking or unhooking it from the battery. I was stunned, and my ears were ringing from the sound, but through my fear and shock, I was also humbled by the awesome power of what I had just witnessed. I marveled at the amount of hydrogen and oxygen gas that came off basically one drop of water, and putting it to work for my entertainment "later on" made it all that more interesting. Making hydrogen and oxygen gas from the electrolysis of water really can get a young mind racing and might set that young mind on a career path to much-needed engineering or scientific trades.

Electrolysis of water to gas is basically a method of separating the elements of water by pushing an electric current through it. Electrolysis isn't the most efficient way to obtain hydrogen, but it is one of the easiest and cheapest ways to make home-brewed hydrogen. Hydrogen is the most abundant element in the universe. Hydrogen is a powerful fuel, and making loud noises in the name of science is a blast.

Direct Current Events

As everyone knows, a water molecule is formed by two elements, which are two positive hydrogen ions and one negative oxygen ion. Electrolyzers are fairly simple technologies in which water is mixed with a conductive electrolyte, and electric current is made to flow past at least one pair of electrodes. One electrode is called the *anode*, and it is the positive side of the electrolyzer, whereas the negative side is called the *cathode*. We all know that opposites attract, and in the electrolyzer, the same is true, so the positive hydrogen ions are attracted to the cathode of the electrolyzer, and the negative oxygen ions are drawn to the anode. Electricity causes the water molecules to split into hydrogen and oxygen gases, which bubble out of the solution on the surfaces of the cathode and anode to eventually become buoyant enough to rise to the water's surface. To make the water better at having electric current pass through it, we're going to need to add what is called an *electrolyte* to it, which is an acid or a base (salt). For our cathode and anode, we're going to use stainless steel because most other metals' electrons could migrate and disrupt the solution. When metals migrate in electrolysis on purpose, it's called *electroplating*. Lower-value metals are often plated with higher-value metals in electroplating; for example, jewelry, metal car bumpers, and even copper pennies are now electroplated. We then add a direct current (dc) of low voltage electricity to the cathode and anode. The current flows through the water, which then separates (splits) into its basic elements. The direct current pulling at the hydrogen's and oxygen's protons is what pulls them apart, making the gas. To make the direct current, we're going to need a 12-volt battery and a solar panel made to trickle-charge the 12-volt battery.

Got Gas?

Once we start producing gas, we're going to need to capture and then ignite the hydrogen and oxygen gases as they exit the water and collect on the surface. To capture the hydrogen and oxygen gases, we're going to use dish soap to catch the gases and form bubbles as they break the surface of the water and try to escape. We'll let the hydrogen and oxygen gas bubbles continue to collect and make a head of foam on the water's surface like a draft beer. When the foam is about an inch thick, we'll ignite the foam with a piezoelectric igniter that generates a nice spark that makes the explosive foam produce a bright flash and a pop (like a firecracker). A piezoelectric igniter works by making a high-energy spark. When you apply pressure, you get a charge separation within the piezo crystal that makes a voltage across the crystal that is extremely high, creating a spark. For example, in a BBQ lighter, the popping noise you hear is a little spring-loaded hammer hitting a crystal or ceramic material that is producing thousands of volts across the face of the crystal. The popping energy to the ceramic produces a voltage when it is hit. The piezoelectric effect is similar to the triboluminescent effect, which is the brief spark (light) produced in your mouth when you crunch on a wintergreen candy. Try it in the dark, looking in a mirror. It's really pretty neat.

WHAT YOU'LL NEED

- Liquid dish soap
- Table salt
- Water
- A recycled 2-liter pop bottle
- A razor knife
- Stainless steel utensils (spoon or fork)
- A 12-volt rechargeable battery
- Insulated electrical wire (2 yards or 2 meters length)
- Wire cutter and stripper
- Electrical tape
- Solar-powered 12-volt battery charger
- A piezoelectric BBQ igniter with an electrical wire attached to it or a conventional flame lighter that has an extended tip like that used for lighting BBQ grills
- A 30-amp on-and-off switch
- A plastic or wooden spoon
- Safety goggles (Figure 10-20)

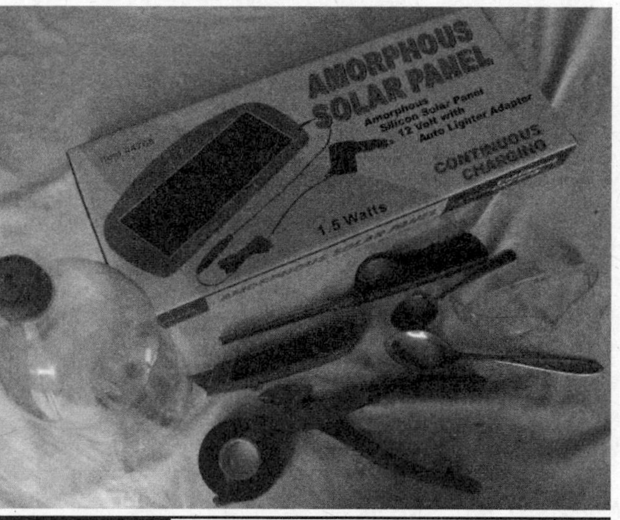

Figure 10-20

TIP We're just making a small amount of gas to make our popping noise and flash, but nevertheless, we still have an open flame and a potentially loud noise, so do not allow young children to play with the hydrogen firecracker under any circumstances. Children cannot understand the danger involved and cannot act appropriately in case of emergency. Only use the hydrogen firecracker outdoors in a clear area away from houses, dry leaves or grass, and flammable materials. Never ignite the hydrogen gas (foam bubbles) in another container, especially a glass or metal container.

Let's Start

1. Cut off the top of the 2-liter plastic pop bottle with the razor knife at the widest point.
2. Now cut two slits across from one another about two fingers' width below the top edge of the cut bottle top (Figure 10-21).
3. Slide the top of the stainless steel utensils (spoon or fork) from the inside of the cut bottle's slits, letting the tops protrude outside the slits on the bottle (Figure 10-22).
4. Cut two pieces of wire about an arm's length, and strip all four ends of the wire. Also add the switch to the middle of one of the wires.

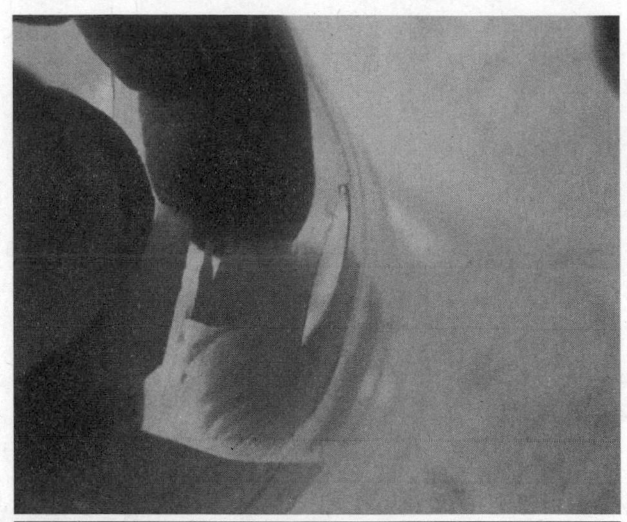

Figure 10-21

Figure 10-22

5. Twist one end of each of the stripped wires around the top of the stainless steel utensils protruding from the slits. And tape them securely together with electrical tape, or you can use battery clips.
6. Now add tap water to just below the slits cut in the bottle.
7. Add several small drops of liquid dish soap to the water in the bottle.
8. Find a sunny spot, and place the 12-volt battery in it. Now hook up your solar-powered charger (follow the manufacturer's instructions) to the battery.
9. Now wire the hydrogen firecracker's wires to the battery (making sure that the switch is off).
10. Once wired, turn on the switch (make sure that it's on), and see if you start making bubbles (dissolved minerals in the water may act as an electrolyte but may decompose quickly). Add a pinch of table salt to the hydrogen firecracker, and stir it slowly with a plastic or wooden spoon until dissolved. Keep adding pinches of salt and dissolving it into your soapy water until you see bubbles start to come to the surface of the water (Figure 10-23).

TIP **Don't add too much salt. It'll make the battery weaken too fast and may make the water heat and make less gas.**

11. When you get about ½ inch of foam on the surface of the water, turn the switch to off.
12. Now, with your goggles on and your arm stretched out, take your source of ignition and let the end of it just barely touch the gaseous foam. Give it a click. POP!
13. Turn the switch back on, and do it again and again by repeating steps 11 and 12. Add water as needed, and give the solar power charger enough time to charge your battery between pops.

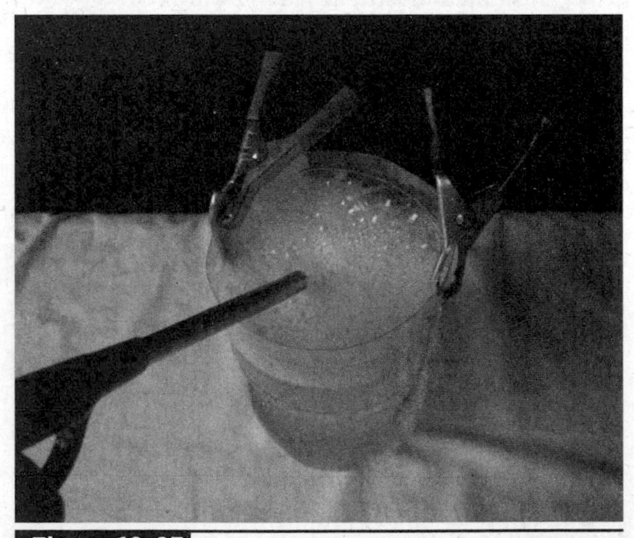

Figure 10-23

TIP Table salt is sodium chloride, and in your hydrogen firecracker, the salt is going to give off a minute amount of chlorine gas, and your salt is going to turn into lye (sodium hydroxide). Electrolysis of sodium chloride is a process industries use to make chemicals for such things as drain cleaner and chlorine bleach.

Project 77
Build a Wind-Powered Scarecrow

We've all seen a set of old clothes stuffed with straw with a pole up its back hanging out in a garden. Sometimes crows even land on them to look out over the garden before they set about their digging. We've even seen those inflatable eyeballs, rubber snakes, and owl decoys. Regardless of the scarecrow type, an effective scarecrow must have motion and some noise. We need to make a scarecrow that moves itself, that has some disruptive motion, that reflects and moves light, and that makes a little noise when there is a breeze.

WHAT YOU'LL NEED

- Unpainted aluminum flashing, about ½ foot (⅓ meter) in width and 1 yard (1 meter) in length
- Rope
- Swivels
- Tin snips
- A razor knife
- A drill
- Drill bits
- Eye protection
- Cut-resistant gloves
- Metal yard or meter stick (Figure 10-24)

Figure 10-24

TIP Always wear eye and hand protection when cutting flashing. To cut aluminum flashing easily, use a razor knife and a straight edge to guide the razor's tip to deeply score where you want to cut the flashing. Once the flashing is scored, carefully bend it with the scored line in the middle of your bend, and it'll split apart as the pressure increases.

TIP This project has sharp edges at all times, so keep it out of reach of people and pets and away from things that it could damage.

Let's Start

1. Cut the sheet metal in half, making two pieces that are the same length (Figure 10-25).
2. Now cut those two sheets from corner to corner, making four short triangles with long edges (Figure 10-26).
3. Drill holes with a drill bit that's slightly larger than the rope you're using for the project. The holes need to be drilled a couple fingers' width from the edges of the triangles' long-side corners.
4. Cut a piece of rope a little longer than the longest edge of the triangles (Figure 10-27).

Figure 10-25

Figure 10-26

5. Lay out the narrowest corners one on top of the other, and thread the rope through all four of them while making a permanent knot in the rope on the other side of the four triangles (Figure 10-28).

Figure 10-27

Figure 10-28

6. Now take the other corners with the holes, fold them over slightly, and thread the rope through the bottom side of each of the triangles, making a sort of pinwheel-looking thing.

NOTE Larger diameters turn slower but catch lighter winds, whereas smaller diameters need stronger winds to turn (Figure 10-29).

7. Tie a permanent knot in the rope on top of the four triangles to keep the desired shape. Use

Figure 10-29

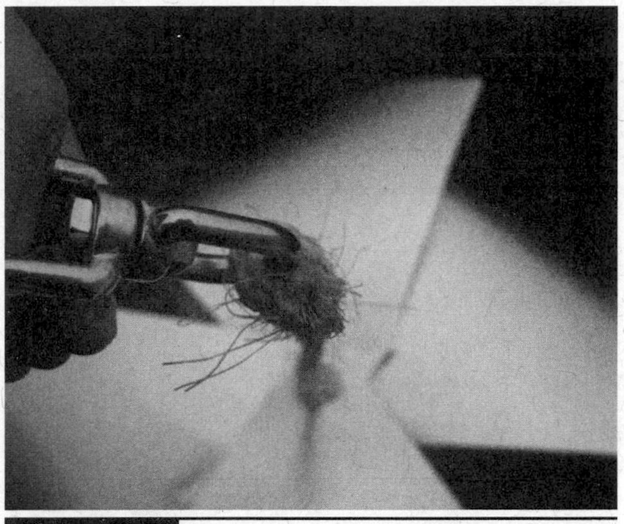
Figure 10-30

duct tape to keep the blades separated, if needed.

8. Add a swivel to the remaining part of the rope (Figure 10-30).
9. Find a location where you want to hang the scarecrow (such as a tree or a hangman's post).
10. Attach another swivel to the area where you want to hang the unit, add some rope to it with a good bit of slack, and tie or fasten it well (Figure 10-31).

TIP **Do this next step on a wind-free day, please.**

11. Then hang the scarecrow by its own swivel on the same rope.
12. Wait for some wind, and enjoy.

Figure 10-31

CHAPTER 11

Making Money with the Items You've Made and Sources

So you have an idea to take some of the things you've made and try to turn them into a source of income. Competing with nonnatural or nongreen types of products does help to open the doors to potential sales because there is an obvious need for these kinds of products. Much like an organic farmer, you want to sell user- and earth-friendly products. You need to have a plan to make, market, and sustain the venture before you invest your money and time. Set goals, and make them work for you.

What's the Plan, Stan?

Develop your marketing plan, so let's start with who would want to buy the types of products you make, and how can you reach these people? Once you know your market, you often have to make your product fit that market. So who makes up the market?

Who Else Is Out There?

Competition is critical to marketing as well as to helping you find your market. Competition is any other product that someone may buy instead of the one you are selling. You must view competition as a tool that you use to make the market work for you. The edge of a green product compared with a nongreen or overly priced green, natural, or even conventional product is that your product is perceived as different—"better" than all the others in function, price, and customer satisfaction. Give potential customers a reason to buy your products over any others, such as a lower price, superior quality, more natural, healthier, and safer to use. The product's "greenness" plays an important role in this, too.

Help Them Find You

Attracting customers involves some kind of advertising, publicity, and personal contact. This is like selling people on the idea of what this product will do for them. A salon doesn't sell haircuts so much as it sells style, appeal, or vanity. Green, natural, and safer products are the buzzwords, so this is the theme of your promotion.

Word of Mouth

This type of promotion is the most valuable for a small business because it's free, simple, and works.

Business Cards

A professionally printed card is an asset. Include your name, address, phone number, and the products you provide.

Brochures

A brochure can be a big plus for helping to impress your buyers.

Direct Mail

You have to know your market and have a mailing list. Keep track of customers' addresses.

Newspapers

Contact your local editors about doing a news story or feature story on your new business. A news story would be done on the fact that you're a new green business, whereas a feature story needs some kind of human-interest angle, so add a personal reason why you're green.

How to Find Your Market

Beat the pavement to see if your stuff is good enough to sell side by side with others like you. It's also important to find the right venue for your work. Do your homework to find out who typically buys products like you have, and see what prices they're bringing. Start small, and try to establishing your reputation. Build a following, and keep your expenses to a minimum. Don't invest a lot of money until you're sure that it's going to work.

The Price Is Right

While cheaper goods are easier to sell, you've got to sell a lot more of them to cover your costs and make a profit. Therefore, establish a fair price that you can live with, and keep your costs low as you start out. Don't be disappointed if you don't make a fortune your first time out.

Up and Up

Look at the local laws to see what you may need to sell your goods. There could be insurance concerns, health issues, and tax liabilities that you may not be aware of. List your ingredients and how to use your goods properly so that people are comfortable with you and what you're doing. If you are thinking about starting a homemade food product business in your home, even on a small scale, then there are many things that you will need to take into consideration before you start selling your product. This is especially true if you have decided to sell homemade products of any sort. Many people who decide to enter into a homemade food product business think that if they keep their operation small, then they don't have to worry about regulations that govern larger food business. Unfortunately, this is wrong.

Places to Sell Homemade Products

The first places to go for hands-on sales are your local farmers' markets, craft fairs, and swap meets. Just be prepared to spend many hours sitting and waiting for your sales to happen. The bonuses in face-to-face sales are social interactions, personal feedback on your product, and coming up with more ideas. If you enjoy people, then direct sales are for you.

Wholesaling your items may offer you freedom from the responsibilities of direct sales, but more often than not it takes more money to be a wholesaler and lowers your net profit. The advantages of wholesaling include that you can have more outlets for your items so that you can turn over greater sales volumes and make more money, if managed correctly.

The best way to sell handmade products is online. First, you'll want to think through whether you want to sell on a third-party site (i.e., a site that's owned by someone else, such as eBay) or your own Web site. There are pros and cons to consider with each approach. Selling crafts online on a third-party site usually allows you to set up your online store quickly and easily, and you can do it with little or no need to understand HTML or Web design concepts.

Third-Party Sites

Third-party sites are full of people who seem to have success selling, but many feel that they struggle with the "garage sale" mind-set. It can be difficult to sell handmade products for a reasonable price there because most people who go to third-party online sites are looking for bargains.

Having Your Own Site

If you are willing to put in the time to create your own site, this is good, but it can be a fairly expensive option. You'll need to invest more time upfront in designing and building your online store than you would if you sold through a third-party site. However, you will benefit from the fact that you own and control the site, and you will not be subjected to the regulations and changes of a third-party site, which could affect your business. You'll have full control over the policies and direction of your online craft store if it is your own site.

One reason people choose to sell on a third-party site instead of setting up their own Web site is that they believe they will automatically, with little or no effort, have customers at their online store because third-party sites, such as eBay, already have plenty of traffic. However, third-party sites also have plenty of sellers who are all competing to get people to their stores. Thus, to create a successful online business, you'll need to invest time to promote your online store whether you sell on a third-party site or your own site.

There are plenty of ways to promote a Web site, and they all require an investment of your time. If your plan is to create an online business, owning your own site is the way to go. There's a lot to be said for spending your time and effort promoting your own site and not that of a third party. You'll be fully in control, and once customers get to your site, you will not be competing with thousands of other sellers, as on third-party sites.

Sources

- *Earth 911.com*—www.earth911.com. Earth911.com is your one-stop shop for all you need to know about reducing your impact, reusing what you've got, and recycling your trash. With years of experience and industry knowledge, this site has the background and resources you need to stay plugged into the green scene.
- *Resource Recycling Systems*—www.recycle.com. Resource Recycling Systems is a nationally recognized environmental consulting and engineering firm that has worked for 25 years to motivate groups and individuals to build a sustainable and waste-free future.
- *Recycle City*—www.epa.gov/ecyclecity. At Recycle City, there's lots to do and people and places to visit, plus plenty of ways to explore how the city's residents recycle, reduce, and reuse waste.
- *Planet Green Recycling*: How to Go Green—www.planetgreen.com.
- *Care2 Healthy & Green Living*—www.care2.com/greenliving/make-your-own-non-toxic-cleaning-kit.html.
- *Nontoxic Home Care*—www.eartheasy.com/live_nontoxic_solutions.htm.
- *Green Home Living*—www.greenhome.com/info/magazine/001/makemyself.html.

- *Green Cleaning Recipes*—www.thedailygreen.com.
- *Greener Choices*—www.greenerchoices.org/products.cfm?product=greencleaning.
- *Natural Pest Control*—www.eartheasy.com/grow_nat_pest_cntrl.htm.
- *Make Your Own Insecticides with These Easy Recipes*—www.pestcontrol.about.com/.../easy-to-make-homemade-insecticides.htm.
- *Easily Custom Make Your Own Insect Repellent*—www.care2.com/.../custom-made-insect-repellent.html.
- *Make Your Own Weed Killer*—www.frugalliving.about.com/b/.../make-your-own-weed-killer.htm.
- *Grandma's Recipe for Fast Weed Control*—www.garden-counselor-lawn-care.com/vinegar-weed-killer.html.
- *Organic Weed Control*—www.landscapingabout.com/.../weedsdiseases/Chemical_Organic_Weed_Control.htm.
- *Natural Weed Killer Recipe*—www.keeperofthehome.org/.../natural-weed-killer-recipe.htm.
- *Soap Making Supplies*—www.soapcrafters.com.

Index

C